CLEAN SOLAR POWER

The Energy Prime Source, and Earth's Rescuer

Omid Malaie

"If you want to find the universe's secrets, think in terms of energy, frequency, and vibration."

NIKOLA TESLA

CONTENTS

PREFACE

"In my view, solar energy is a blessing to increase human survival and postpone its extinction because the sun is the source of all energies (except nuclear and geothermal). Humans use its types indirectly and latterly as fossil fuels. Today, due to climate changes and the lack of fossil fuels, humanity seeks to develop technologies for the direct and immediate use of abundant and renewable solar energy much more than in the past." Omid Malaie In the vast expanse of our solar system, the sun stands as the unparalleled source and catalyst for nearly all forms of energy on Earth, with the exception of nuclear and geothermal sources. Beyond merely illuminating our days, the sun's influence extends to every facet of our existence.

Comprising a staggering 99.86% of the solar system's mass and volume, the sun plays a role far more profound than the limits of human imagination can grasp. Recent statistical analyses underscore the escalating growth and significance of harnessing solar energy, eclipsing other energy sources in relevance.

This book delves into the pivotal questions surrounding the sun and solar energy's vital role in our solar system and on Earth. Can we, with the aid of this celestial powerhouse, affordably and sustainably supply the entire global population with the energy needed for their daily pursuits? How does solar energy address the electricity needs of over a billion people who lack access to it? What opportunities and challenges lie ahead? And, perhaps most pressing, what impact will the abundance of free solar energy have on human life, fossil fuels, pollution, and the looming specter of global warming?

Join us on a journey to explore the promises, possibilities, and potential obstacles that solar energy presents a journey that may redefine the trajectory of our energy landscape.

Energy, water, and food are the three main elements that ensure human survival and are interdependent. For example, if we have enough energy, we can provide more water and food. Since every tool and technology needs energy to do any work, knowing the energy and its types, its correct use and application, and the correct decisions in this field is essential. According to the analysis and data of the International Energy Agency report, solar energy will reach first place as the most-used source of energy by 2027. In addition, the sun is almost the primary source of most other energies, whose detailed investigation leads to a deeper understanding that is more applicable to developing new technologies. In this chapter, the types of energy and their classification have been reviewed in summary form. Later, the primary sources of generation and conversion of energy and solar energy are explained in detail. Finally, energy storage, transmission, and distribution, which are essential parts of new technologies, are briefly described.

TYPES OF ENERGY AND
ITS SOURCES

Energy is the ability to work or bring about change. It exists in many forms, such as potential, kinetic, chemical, nuclear, and thermal. Energy can be transferred between objects and transformed into other forms but cannot be created or destroyed. If we accept that existence consists of "matter" and "energy," then understanding matter seems relatively more straightforward because energy is one of the deep and complex concepts in all sciences. Unlike matter, energy cannot be smelled, seen, or touched most of the time. Although most of us are familiar with energy, it is difficult to define because many forms of energy are not a smelly, visible, and tangible "thing" (in terms of having mass). On the other hand, the relationship between matter and energy makes this definition more difficult.

Perhaps, considering the scope of the book and the world of technology, it is better to use the concept of "work" to define energy, in which case it can be said that energy is equivalent to "the ability to do work"; which is one of the most straightforward definitions of energy in physics. This definition means that we need energy to do anything. It should be noted that work is not a form of energy but a method of transferring energy from one point to another or converting it from one type to another. In this definition, energy is closely related to the concept of "force." In classical physics, force refers to the effect that objects can create and change in dynamics. Force is a vector quantity, meaning it has both magnitude and direction. According to a more general and, of course, more modern definition, there are four main types of interaction between bodies, which, in order of strength from high to low, include the following: strong nuclear force, electromagnetic force, weak nuclear force, and gravitational force.

Energy appears in various forms, which include kinetic energy, thermal energy, electromagnetic radiation energy, gravitational potential energy, elasticity, electrical, chemical, nuclear, and ionization energy.

Today, many energy sources are available for harnessing and mining, each with advantages and disadvantages. These sources include the sun, wind, water, fossil fuels such as coal, oil, and gas, radioactive elements such as uranium, geothermal and biomass, tides, and oceans.

Understanding energy allows us to harness it responsibly for transportation, electricity, heating, and industrial applications that power economic growth and development. Optimizing energy efficiency and adopting green technologies are crucial to ensuring access to clean, affordable energy for the world's growing population. Managing energy consumption wisely today secures our energy supplies for the future.

CLASSIFICATION OF DIFFERENT TYPES OF ENERGY

There are different classifications and separations for energy based on various aspects. They can be separated in terms of presence, environmental, sustainability, renewable, green, and clean, and these divisions may have commonalities, similarities, and differences. Not all renewable energy is sustainable energy, and vice versa.

Excessive pollution of the environment due to the use of fossil fuels is one of the enormous crises that can be solved by introducing new energy. Energy use is considered sustainable if it meets the needs of the present without compromising the needs of future generations. Definitions of sustainable energy typically include environmental aspects such as greenhouse gas emissions and social and economic aspects such as energy scarcity. Burning fossil fuels and biomass[1] is one of the leading causes of air pollution, which causes thousands and even millions of people to die yearly. There are still challenges to overcome regarding cost, infrastructure, and politics to develop sustainable energy.

The useful energy obtained from renewable sources that do not involve carbon and are naturally charged and stored by nature in a short period, such as the sun, wind, rain, tides, waves, and the heat of the earth's core, energy is called renewable. On the other hand, non-renewable energies such as fossil fuels take a long time to produce and pollute the environment due to the presence of carbon. These energy sources lead to environmental degradation and climate change due to their carbon-intensive nature and extraction processes. Many renewable energies are classified as sustainable, but some, such as biomass, are not sustainable because they pollute the environment.

Green energy usually refers to renewable energy sources that have the lowest environmental impact and do not result in carbon production. In the end, even other renewable energies, in addition to biomass, have losses to the environment. Green energy includes solar, wind, hydro, and geothermal because they produce clean electricity and have a lower carbon footprint than conventional energy sources. It is important to note that there can be overlap and different perspectives in these classifications, as the assessment of energy resources may depend on specific contexts, technologies, and socio-economic factors.

The benefits of sustainable energy go beyond its renewables. Sustainable energy offers significant environmental benefits, such as reduced greenhouse gas emissions and air pollution associated with traditional energy sources. In addition, sustainable energy sources are often more and better distributed, promoting energy independence and reducing the vulnerability of centralized power systems. Advances in technology, infrastructure, and policy support are essential to realize the potential of sustainable energy fully. Continuous investment in research and development is necessary to improve efficiency, reduce costs, and integrate different sustainable energy sources into the existing energy grid. Sustainable energy encompasses a broader scope, including environmental aspects, socio-economic impacts, and long-term resource management.

- Sustainable energy considers the entire life cycle of energy systems, including resource extraction, production, distribution, and end-of-life disposal. The program promotes holistic energy production, consumption, and management strategies to ensure a sustainable energy future.

- While green energy primarily focuses on reducing carbon emissions, sustainable energy considers other aspects of sustainability, such as energy efficiency, biodiversity conservation, social equity, and economic viability. Sustainable energy goes beyond renewable energy sources and includes energy efficiency measures, energy-saving practices, and integration of energy systems that promote optimal use of resources and reduce losses and waste.

- Green energy is often marketed as a consumer choice or a specific product, such as purchasing green energy credits or installing solar panels. On the other hand, sustainable energy encompasses a broader dimension, including vision, policy frameworks, technological developments, and systemic changes in the energy sector.

Green energy can be considered a subset of sustainable energy focusing on carbon reduction and low or zero emissions. *Sustainable energy* is a multidimensional concept that integrates environmental, social, and economic considerations in the planning and managing energy systems with the goal of long-term sustainability and resilience.

THE PRIMARY SOURCES OF ENERGY EXTRACTION AND THE CONVERSION OF DIFFERENT TYPES OF ENERGY INTO EACH OTHER

Energy conversion is fundamental in shaping our modern world, advancing and driving our daily activities. Energy conversion has been happening since billions of years ago, and today, by knowing these old natural processes, humans act in energy conversion for their own unique and new purposes. The ability to efficiently convert one type of energy into another form is essential in many areas, including transportation, industry, and our personal lives. This process is necessary for several reasons. One of the main reasons lies in the finite nature of our energy resources. Energy conversion enables us to take advantage of different forms of energy and harness their potential effectively. Renewable energy sources have vast reserves that guarantee clean and long-term energy supply. By effectively converting these resources, we can reduce our carbon footprint, fight climate change, and protect our planet for future generations. Energy conversion also has tangible benefits for transportation systems. The industry also relies heavily on energy conversion processes. Factories and facilities require significant amounts of energy to power their operations. Efficient energy conversion allows industries to minimize waste, optimize production, and reduce energy consumption, contributing to economic and environmental sustainability. Using cleaner energy sources in industrial processes reduces harmful emissions and ensures the well-being of surrounding communities. In addition, energy conversion has a significant impact on the daily life of humans. Energy conversion makes Many conveniences possible, from charging electronic devices to heating or cooling homes. Energy conversion technologies, such as solar panels or heat pumps, enable homeowners to efficiently generate electricity or use natural resources such as geothermal heat. These procedure processes reduce energy costs and allow people to participate actively in the transition to a greener future.

As a result, energy conversion has an essential place in our society and global development. By converting and optimizing different forms of energy, we can effectively address the challenges posed by limited resources, climate change, and unsustainable practices.

This process is crucial in achieving a stable, clean, and prosperous future for users and future generations. This section examines the primary sources of energy generation and the conversion of different types of these energies into each other.

The primary sources of energy extraction

The sun is the ultimate source of all forms of energy on Earth, except for nuclear and geothermal energy. Solar energy is harnessed on Earth through various methods. For example, solar panels, which consist of photovoltaic cells, absorb sunlight and convert it into electrical energy through the photovoltaic effect. This electrical energy can be used in different ways. In addition to the direct harnessing of solar energy through approaches like photovoltaics, concentrated solar power for thermal applications, lighting requirements, photochemical systems, and prudent solar architectural layouts amongst others, indirect derivations of solar energy also contribute to renewable generation through modalities including wind energy (a result of uneven solar heating of the atmosphere), hydropower (solar evaporation leading to rainfall feeding dams), biomass production through photosynthesis, and ocean energy from temperature gradients and waves. It provides other renewable energy and has already provided energy from non-renewable sources. For example, wind energy is produced when sunlight heats different parts of the Earth unevenly, causing temperature changes and thus air pressure. These changes lead to the movement of air and the creation of wind, which can be used through wind turbines to generate electricity. Tidal energy occurs due to the effects of the moon's gravitational forces and the sun on the Earth. The sun's impact on tidal energy is indirect through its gravitational pull, affecting tides' height and timing. The time required to transfer and create tidal and wind energy is approximately a few minutes to a few hours.

Similarly, hydroelectric power relies on solar energy. The sun evaporates water from oceans, lakes, and rivers, forming clouds that eventually lead to rain or snow. This accumulated water flows through rivers and streams, driving turbines and generating electricity as it passes through hydroelectric plants. This process takes several weeks to several months. Even fossil fuels such as coal, oil, and natural gas ultimately derive from solar energy. Millions of years ago, plants grew by absorbing sunlight through photosynthesis. When these plants died and decomposed, they formed layers of organic matter that, over time, turned into fossil fuels through heat and pressure deep in the Earth's crust.

There are only two sources of nuclear and geothermal energy; their primary source is not the sun star. Nuclear energy is produced from radioactive elements such as uranium. The presence of radioactive elements such as uranium on Earth can be explained by the processes that occurred during the solar system's formation. These elements were mainly produced during the supernova explosion of massive stars about 6 to 7 billion years ago. This process, known as nucleosynthesis, involves the fusion of lighter elements, such as hydrogen, in the cores of stars to form heavier elements, such as uranium, iron, and other elements. Another energy of the Earth, which existed at the same time as the formation of the sun and the solar system, is geothermal energy. Earth's core is hot for two reasons: residual heat from the formation of the planet and the decay of radioactive elements inside it. When Earth formed approximately 4.6 billion years ago, it underwent accretion, in which smaller planetary bodies collided and merged, releasing significant amounts of energy. This process produced heat and caused the Earth to melt partially. Over time, as the planet continued to accrete and gravitational forces compressed the core, the molten material solidified but retained much of its initial heat. Another warming of the Earth's core is the decay of radioactive elements. Earth's core contains small amounts of radioactive elements such as uranium, thorium, and potassium. These elements were present early in the formation of the Earth and are still subject to radioactive decay. During this process, the atomic nuclei of these elements break spontaneously, and some of the initial mass is released as enormous heat energy.

Figure 1 depicts how the energies are generated and transferred to the Earth along with the required time and periods.

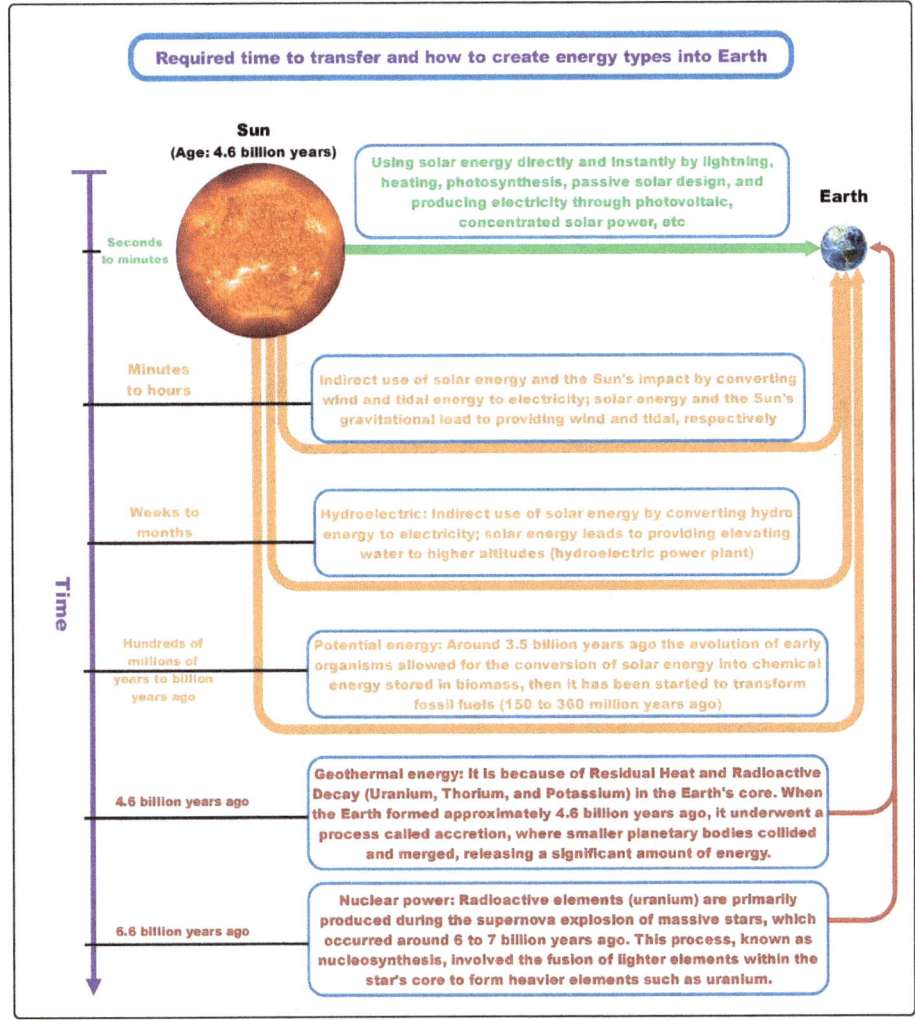

Figure 1: Creation and transferring energy to the earth

Converting different types of energy into each other

In this section, the conversion of different types of energy to each other has been investigated, and all energies originate from the sun, except for nuclear and geothermal energy. In the universe, energy exists in different forms, shown in Figure 2, their conversion to each other. Various forms of energy are constantly transforming into each other. However, the energy source can be considered the sun, and the final destination of all of them is thermal energy. Usually, the chemical energy stored in fossil fuels is converted into thermal energy by burning, and the efficiency of these combustions is almost 90%. Thermal energy can be converted into mechanical energy using heat engines. Heat engines have an energy conversion efficiency of up to 60%. Carnot's efficiency constraint ultimately limits their performance, which is the principle that most current cars and heat engines work on. Mechanical energy can be converted into electricity using electric generators with an efficiency of up to 99%. Currently, most of the world's electricity is produced by a turbo-generator connected to a steam turbine (Rankin cycle), and the primary source of steam turbine energy is mostly coal. During the process of generating electricity from fossil fuels, at least 50% of the primary chemical energy contained in these fuels is lost.

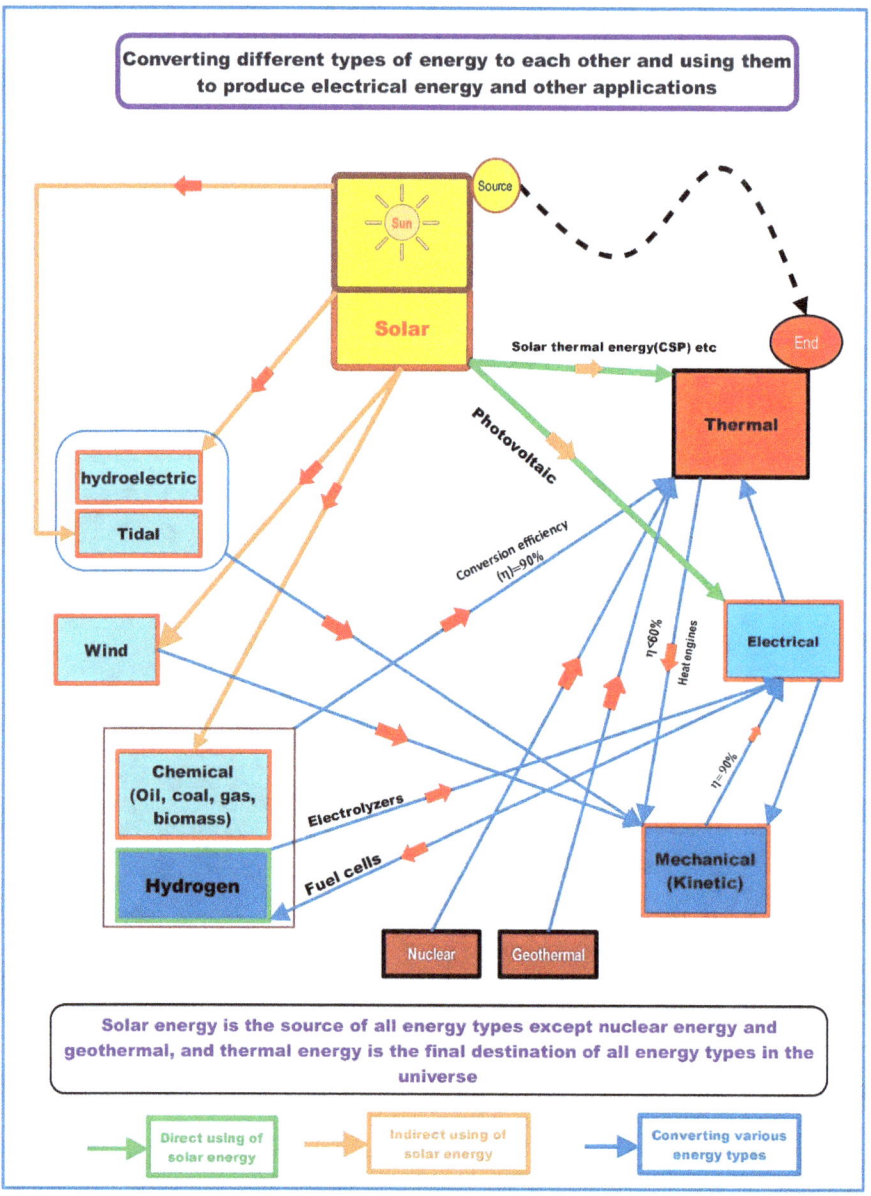

Figure 2: Converting different types of energy to each other and using them to produce electrical energy and other applications (one of the main goals of energy conversion on a large scale is to produce electricity, but it is not always the main goal)

Chemical energy can be directly converted into electricity using a fuel cell. The most common fuel used in fuel cell technology is hydrogen. The energy conversion efficiency of fuel cells is almost 60%. A regenerative fuel cell can work in both directions and even convert electrical energy into chemical energy, called electrolysis. The typical conversion efficiency of hydrogen electrolyzers is 50-80% [1].

Nuclear power plants release energy as heat during nuclear fission reactions. Water vapor is produced with the help of heat obtained from nuclear fission. The continuation of producing electricity from this water vapor is similar to other electrical energy production cycles, the Rankine cycle, next, with the help of the produced steam, a steam turbine and then an electric generator moves, like most fossil fuel power plants, which supplies the required electricity.

Electricity

Figure 3 shows the share of energy consumption in different years. In 2020, 20.5% of the world's total secondary energy is used as electricity, and this share is still increasing. Figure 3 also shows how electricity is supplied. Electricity is a form of energy that can quickly and cheaply transmit through the national power grid with relatively small and clean losses. The transfer and use of this type of energy are more accessible and safer than other energies, such as natural gas. Electrical energy can be converted using mechanical devices into mechanical, thermal, and chemical energy, and vice versa. This ability to convert into different types of energy has made electricity a widely usable form of energy. Electricity is essential in many aspects of our lives, including lighting, heating, communication systems, electronics, industries, electric transportation, medicine, and many others. This wide range of applications has made electricity a significant energy source in daily life and industry, and its use is increasing daily. It is essential to understand that modern society as we know it would not be possible without electricity. Electricity has been in practical use for over 100 years, and access to electricity greatly determines our standard of living.

According to the latest data from the International Energy Agency, by 2021, approximately 759 million people worldwide will have no access to electricity. This figure shows a significant decrease compared to previous years and progress in expanding the world's access to electricity. Approximately 1.7 billion people worldwide only had access to electricity in 2000. In 2021, about 61% of electricity will be generated using fossil fuels, with coal as the primary contributor because coal emits about twice as much carbon dioxide per kilowatt-hour produced as natural gas; coal-fired power plants are a significant contributor to global warming. Nuclear energy accounts for 8.9% of the world's electricity production, and hydroelectricity has the largest share among renewable energy sources with 15%. Of the total electricity produced, about 40% of electrical energy is used for residential purposes, 47% is used in industry, and 13% is lost in transmission. In recent years, transportation has not played an essential role in electricity consumption, but their contribution is expected to become significant due to the importance of electric vehicles. Figure 4 shows the share of annual electricity generation based on energy sources. Notably, the percentage of electricity production from sustainable energies is growing only for solar and wind energy. Among all of them, the rate of growth of solar energy is the highest, and the share of production from hydropower, nuclear, and coal sources is constantly decreasing.

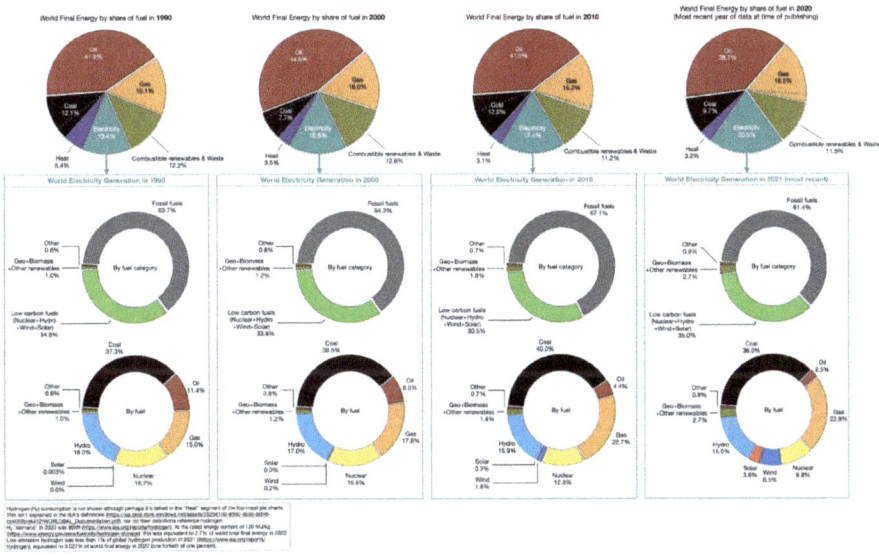

Figure 3: Energy consumption by energy services and energy carriers used to generate electricity for different years [2]

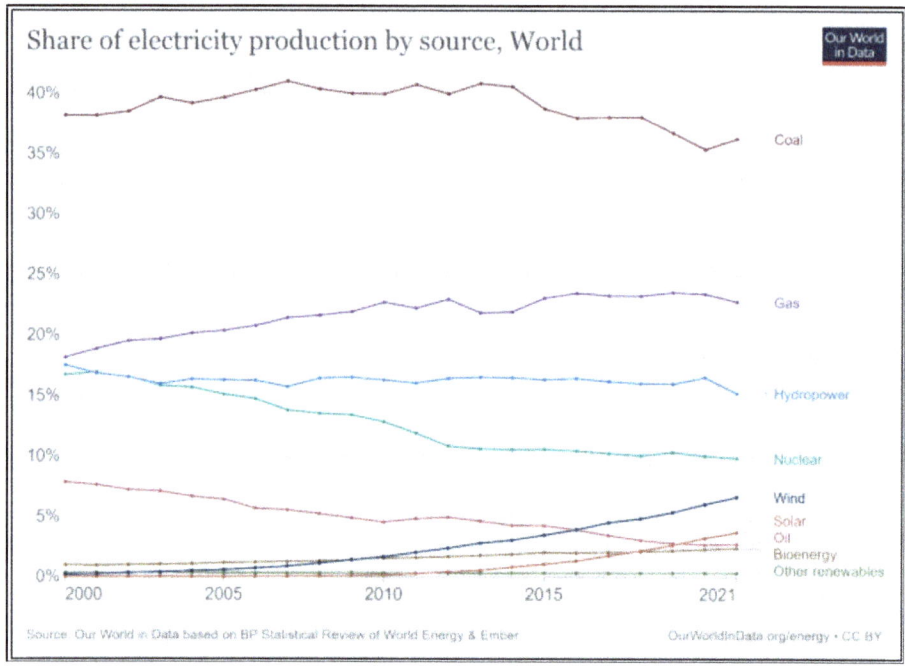

Figure 4: Share of electricity generation based on energy sources

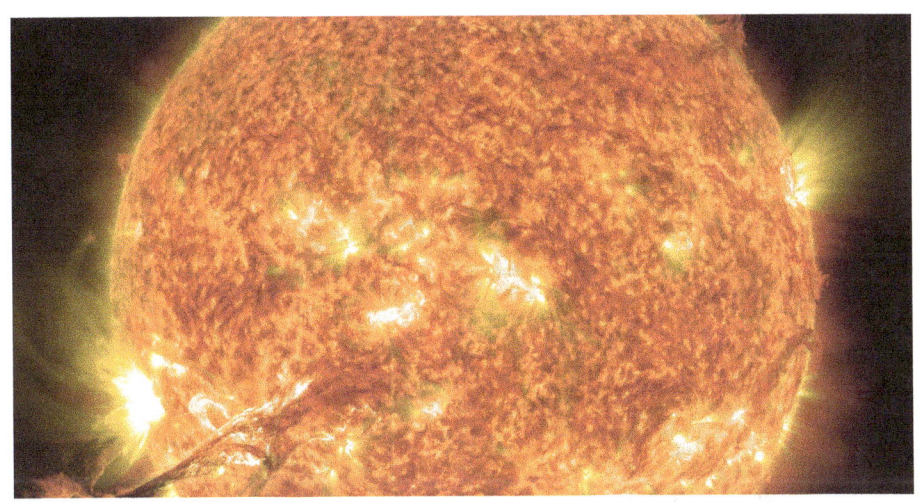

THE SUN AND SOLAR ENERGY

The closest image of the Sun star by the European probe on March 24, 2022 [3] (this photo alone contains 99.86% of the total mass and volume of the solar system!)

Figure 5: A student studying with a candle in an area inaccessible to energy and electricity

Figure 6: Providing solar electricity in a virgin area in Africa

The massive expansion of solar energy use worldwide by the middle of this century will likely be essential to any severe approach and strategy to mitigate adverse climate change. Fortunately, the free and clean source of solar energy has the potential to supply a significant portion of future energy needs. The sun is an abundant and virtually inexhaustible source of energy that has dramatically outstripped our current global energy needs and can supply our energy needs. In recent years, the cost of using solar energy has decreased significantly, and its usage is increasing. Today, solar energy accounts for only about 3.74% of the world's electricity production [4]. The use of non-renewable energy leads to the production of carbon dioxide. Carbon dioxide production imposes a harmful and destructive cost on the environment and humans. To compare the cost of using different energies, if the unfavorable cost of carbon dioxide production is considered, using renewable energies, especially solar energy, is entirely economical [5]. If there is no penalty for carbon dioxide emissions, it will be impossible to increase the capacity of using solar energy to a high level.

Nevertheless, a significant change in government policies is implemented along with the climate challenge. In that case, the use of free, clean, and inexhaustible solar energy will increase at a much faster rate. Unfavorable policies, intermittency, high cost, and scalability are the most significant obstacles and problems related to solar energy and other renewable energies, which are discussed in detail below. Although the growth rate of the use of solar energy is higher than all different energies, these obstacles have still slowed down further progress in this field.

The International Energy Agency has predicted that solar energy will surpass all other sources from 2027 onwards. The agency has indicated that in 2027, 22.22 percent of energy will come from solar photovoltaics alone (excluding other solar technologies). The diagram of this report is shown in Figure 7.

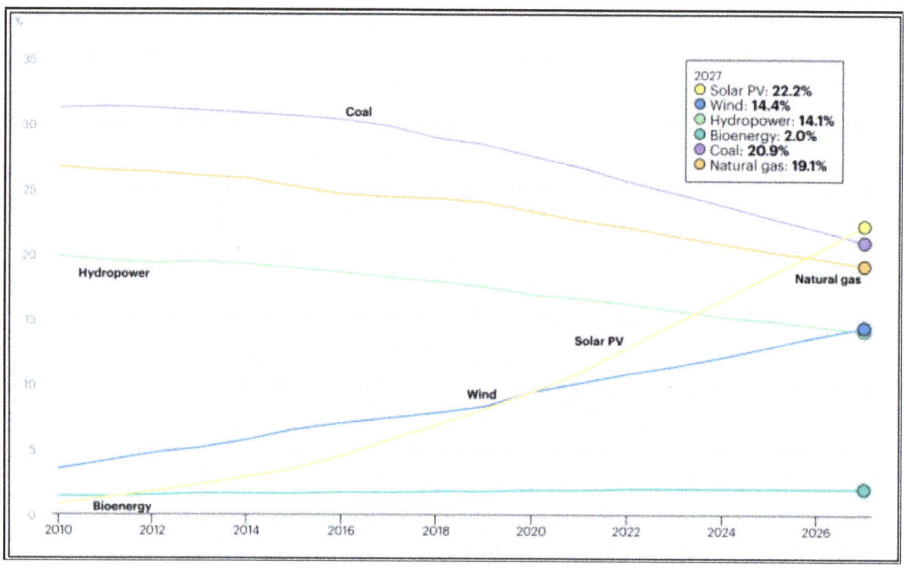

Figure 7: Forecast of the share of energy resources until 2027 by the International Energy Agency (IEA) [6]

Figure 8 compares the growth rate of renewable energy resources in different years based on percentages. The growth rate of solar energy compared to other renewable energy sources has been in the first place for all recent years. It is noteworthy that hydroelectric energy currently provides a large share of energy globally. However, the increase in capacity is facing problems, and in the graph drawn, its growth rate is the lowest. In general, renewable energy and natural gas use has seen high growth in recent years; oil has grown moderately, and coal and nuclear power have decreased. The reason for these changes is the general movement of the global energy mix towards low-carbon and environment-friendly sources.

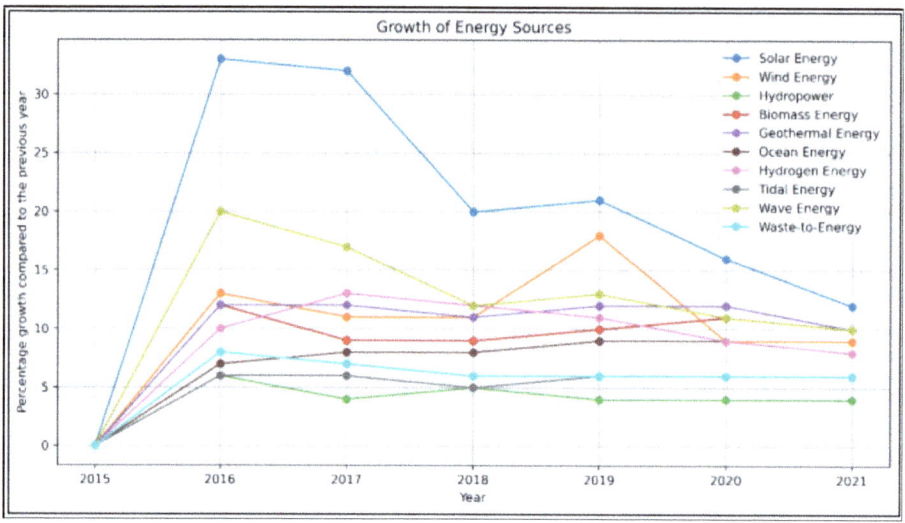

Figure 8: Comparison of the percentage growth rate of energy resources in different years

Solar power generation is one of the few low-carbon energy technologies that has the potential to grow on a vast scale. Consequently, massive global solar generation capacity expansion to the multi-terawatt scale is likely a necessary and feasible strategy to mitigate the risk of adverse climate change. Recent years have seen rapid growth in installed solar generation capacity, significant technological advances, price and performance, and the development of creative business models that have encouraged investment in various residential solar systems. However, further developments are needed to increase the share of solar energy at acceptable costs. Advances in these areas will help efforts to reduce greenhouse gases and develop electrical systems around the world. It also provides light, energy, power, well-being, and comfort to more than one billion people worldwide who live without electricity. Solar energy can be converted into electricity, heat, and chemical energy. The sun is the source of energy and responsible for almost all energy transfer processes that occur on Earth except for nuclear and geothermal energy. Wind is caused by the temperature difference between two areas of the environment caused by the sun's radiation. Waves are created by wind and the gravity of the sun and moon. Clouds and rain are formed by the evaporation of water by sunlight. Therefore, since the sun is almost the only energy source, man must go to a period where he directly uses solar energy to satisfy his energy needs.

In this chapter, the study of solar energy and various methods and systems of its use have been studied. Also, other energy sources such as fossil fuels and other renewable energies are mentioned as necessary because the sun is their primary source. Also, these concepts are essential to understand the chain of conversion of different types of energy to each other and to a comprehensive understanding of the field of energy. Advantages, disadvantages, obstacles, and problems are discussed. The impact of this clean, free, and renewable energy on human life has been discussed in terms of welfare, economy, environment, and pollution, which leads to the correct use of solar energy in a practical way to meet needs. Also, the policymakers' previous and future favorable and unfavorable policies were discussed, and suggestions were made, which can be a good reference for the decision-makers. In this chapter, in the order of basics related to solar energy and how it works, overview and overview of solar energy, review of some technologies of its use such as photovoltaic systems and centralized thermal power plants, various applications of solar energy, development and combination of systems Solar is explained with other energy sources and marketing. Finally, the summary of the presented materials has been examined in the form of a final approach.

The sun and solar radiation

The Sun is the star at the center of our solar system. The star is a giant glowing ball of hot gas composed mainly of hydrogen (about 70 percent) and helium (about 28 percent). The diameter of the Sun is about 1.4 million kilometers, which is about 109 times the diameter of the Earth, and its mass is about 330,000 times that of the Earth. The Sun is vital in sustaining life on Earth, providing heat and light essential for various biological processes and the planet's overall climate. The mass of the Sun is so high that it constitutes 99.86% of the total mass and volume of the solar system. The mass of the Earth is only 0.17% of the mass of this system. Its structure is depicted in Figure 9, and some of its features are mentioned in Table 1.

The surface of the Sun is covered with dark spots called sunspots, which are regions of intense magnetic activity. Above the photosphere (the outermost layer of the Sun), the Sun's atmosphere consists of several layers. The first layer of the Sun's atmosphere is the chromosphere, followed by the transition zone and the outermost layer called the corona. The temperature in the chromosphere (the first layer of the Sun's atmosphere) starts at about 4,500 degrees Celsius (8,132 degrees Fahrenheit). It increases with altitude, rising to millions of degrees Celsius in the transition zone. The Sun's corona extends to millions of kilometers in space, and during a total solar eclipse, it is visible from the Earth as a weak and pearly white halo. The Sun also emits a steady stream of charged particles called the solar wind, which spreads throughout the solar system. Occasionally, the Sun emits massive bursts of plasma and magnetic fields known as solar flares and coronal mass ejections. These can affect Earth's space weather and cause geomagnetic storms.

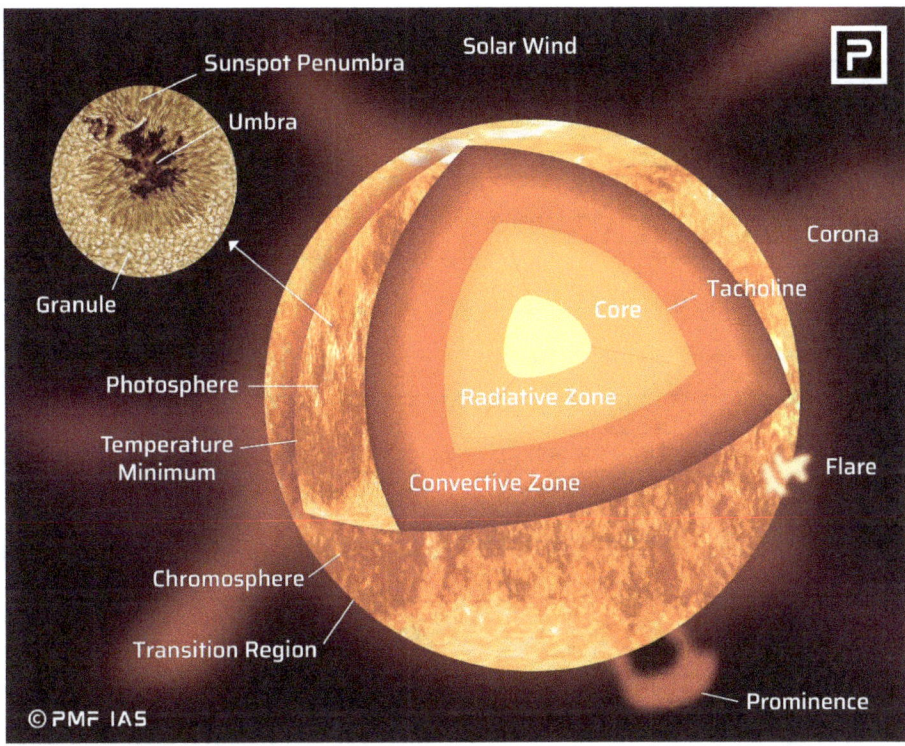

Figure 9: The structure of the Sun [7]

Table1 : Some characteristics of the Sun [7]

Age	4.6 billion years
surface gravity	274 m/s^2 (28 times the gravity of the Earth)
Speed of rotation	7179.73 km/h (Earth's rotational velocity is 1675Km/h)
Period of rotation	25 days 9 h
Rotation	counterclockwise (when viewed from a long way above Earth's north pole)
Composition	98% of the Sun is

	hydrogen & helium
Mean distance from the Earth	149600000km (the astronomic unit, AU)
Diameter	1392000km (109 × that of the Earth)
Volume	1300000 × that of the Earth
Mass	1.99×10^{27}kg (332000 times that of the Earth
Density (at the center)	$> 10^5$kgm^{-3} (over 100 times that of water)
Density (at the surface)	**1410 $^{\text{kgm}^{-3}}$ (1.41 times that of water)**
Pressure (at the center)	over 1 billion atmospheres
Temperature (at the center)	about 15000000 °C
Temperature (at the surface)	5500 °C
Energy radiation	3.8×10^{26}W
The Earth receives	1.7×10^{18}W

The Sun's energy is produced through nuclear fusion, in which hydrogen atoms combine to form helium, releasing large amounts of energy in the form of light and heat in the process. The process of nuclear fusion also occurs in other stars. It means that the fuel used by the Sun is hydrogen. If the solar fuel runs out, life in the solar system will end. However, it takes billions of years to run out of all the Sun's fuel; that is why solar energy is called renewable energy. The fusion process requires very high temperatures and pressure available inside the Sun. The temperature of this star is estimated to be about 15 million degrees Celsius (27 million degrees Fahrenheit), and due to the massive mass of the Sun, the pressure inside is very high. The Sun's core is so dense that radiation cannot travel freely but is constantly absorbed and reflected, so it takes 10,000 to 170,000 years to reach the Sun's surface!

During nuclear fusion in the Sun, mass is converted into energy according to Einstein's famous theory of special relativity. Although a small amount of mass is converted into energy in every nuclear fusion, according to this renowned theory, the energy released is tremendous, even for a tiny mass [2]. For example, 167 grams of a substance equals 11.4 billion kilowatts of energy. This energy is needed for a big country like America, with 340 million people (in August 2023) for one year! Although converting nuclear energy is complex and sensitive, this is impossible. Figure 10 shows a schematic of the nuclear fusion process in the Sun.

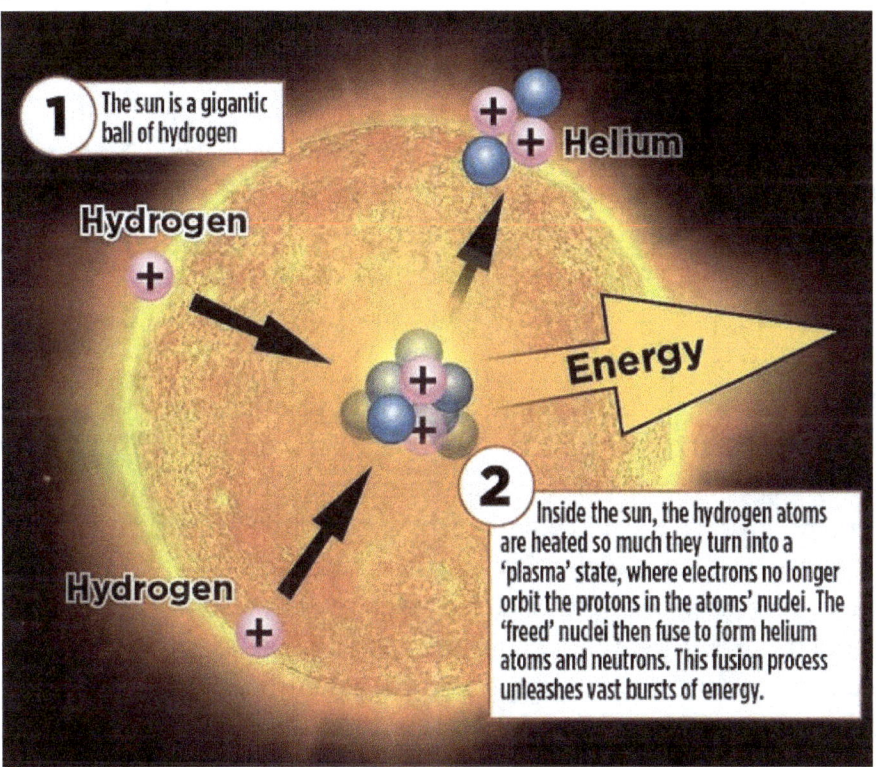

Figure 10: Nuclear fusion inside the sun [10]; Hydrogen atoms are converted into helium; during this process, some of the initial mass is converted into energy and is released as light and heat (electromagnetic waves) [8]

Approximately 4 million tons of mass are converted into energy in the Sun every second. Based on that, the power density in the center of the Sun is estimated to be about 275 watts per cubic meter with theoretical assumptions. The amount of mass that the Sun loses every second is vast, but the total mass loss is minimal compared to the total mass of the Sun. The Sun maintains its mass and energy balance by continuously converting hydrogen into helium in its core.

Neutrinos are produced in the nuclear process in the Sun. Neutrinos are tiny particles with no electrical charge and are ejected from the Sun almost immediately after production. They are highly permeable both in vacuum and matter and have virtually no interaction with matter. Neutrinos in the Sun hardly interact with matter and, therefore, can leave the Sun's core unhindered. Every second, about 6.5×10^{10} neutrinos pass through the Earth per square centimeter, and as a result, they also pass through our bodies. Neutrinos comprise about 2% of the total energy the Sun radiates. The rest of the radiation is released as electromagnetic radiation.

The Sun's outermost layer is called the photosphere, which we see as the surface of the Sun. This part behaves like a black body and is the source of solar radiation hitting Earth. A blackbody is an ideal physical object that absorbs all incoming electromagnetic radiation without reflection or transmission. While the Sun shares several properties with a black body, it does not fit the ideal model perfectly. Real-world factors such as the Sun's constituents, magnetic fields, and atmospheric effects influence the behavior and spectral properties of the Sun. Because of this, its temperature appears yellow, and its average surface temperature is about 5,500 °C (9,932 °F). The solar constant is the total power density of solar radiation at the average distance between the Earth and the Sun in a plane perpendicular to the direction of the Sun, which is approximately 1361 W/m^2.

Stefan Boltzmann's law [3] states that as the temperature of the black body increases, more energy is emitted exponentially and with the fourth power per unit area. That is, the higher the temperature of an object, the greater the intensity of the corresponding radiation and increases to the fourth power. This relationship is essential for understanding various phenomena related to thermal radiation, such as stars' energy output, bodies' heating, and planets' temperature distribution. It is worth noting that while Stefan Boltzmann's law was derived for ideal black bodies, it provides an accurate description of the behavior of real bodies that absorb and emit radiation.

Understanding the Sun's behavior and studying its various phenomena is crucial for space exploration, satellite communications, and predicting space atmospheric events that can affect technology and infrastructure on Earth. Scientists use telescopes, satellites, and other instruments to observe and analyze the Sun's activity, providing valuable insights into the physics of stars and the nature of our universe. Scientists on Earth are also actively working on developing controlled nuclear fusion as a potentially clean and nearly unlimited energy source. However, replicating the conditions of the Sun's core in a controlled manner is an essential scientific and engineering challenge that has yet to be fully overcome.

Solar Energy

One of the carriers of renewable energy is solar energy. Solar energy technology converts sunlight into electricity using photovoltaic cells and concentrated solar power. These methods are explained in the following sections. The technology has become increasingly affordable in recent years, making it an attractive option for homeowners and businesses seeking independence from fossil fuels. If sunlight is directly converted into electricity using devices based on semiconductor materials, it is called photovoltaic. Typical efficiency of solar modules is in the range of 15-20%. Sunlight can also be converted into heat, called solar thermal energy or concentrated solar power. For example, water or another fluid can be heated by sunlight mediated by a black absorbing material. This hot fluid can be used for heating buildings or even cooling. From concentrated solar energy systems, a fluid with a temperature of several hundred degrees is obtained, which is used to generate steam and drive a steam turbine, followed by a generator, to produce electricity. In addition to producing heat and electricity, solar energy can be converted into chemical energy. This technology is called solar fuel. Photovoltaics and regenerative fuel cells can be combined to produce solar fuels. In addition, sunlight can also be directly converted into fuel using photoelectrochemical devices. In the following, photovoltaic modules and concentrated solar power are discussed in detail.

Solar energy is a renewable and abundant source of energy that comes from the sun's rays. It is increasingly attractive as a clean energy solution to combat climate change and reduce dependence on fossil fuels. There are various technologies for using solar energy, which are discussed in the following sections.

Today, solar energy has a relatively small share of electricity generation worldwide. In 2021, this share is about 3.74% of global electricity production. However, the growth rate of this share is significant compared to other sources of electricity production. With this growth rate, it is predicted that this energy will play a vital role in the global energy system by the middle of the century. As several significant barriers to solar energy are removed over the next few decades, the likelihood that this energy will be able to meet critical needs will increase dramatically. This study aims to set up appropriate solar systems, help decision-makers about solar energy in the future, identify barriers that may prevent solar technologies from achieving this potential, and identify applicable public policies that can mitigate current obstacles and problems.

The increasing importance of solar energy is due to the depletion and scarcity of fossil fuels in the future and the profound, long-term threat of global climate change. The emission of carbon dioxide caused by the combustion of fossil fuels causes climate change and global warming. Carbon dioxide is the largest share of greenhouse gases in the atmosphere, increasing ever since the Industrial Revolution. Almost two-thirds of carbon dioxide emissions from fossil fuels are related to electricity generation, heating, and transportation. To mitigate issues associated with climatic and environmental changes, carbon dioxide emissions as a proportion of global energy consumption must be considerably reduced at rates aligned with global economic growth.

Solar energy has the potential to play an essential role in meeting energy needs in the future. The technology of using solar energy to produce electricity with very low carbon dioxide emissions has been developed to an acceptable level, and the use of electricity for industry, transportation, heating, etc., has already become clear. In addition, solar energy sources are abundant, can supply a large part of the world's energy consumption, and exceed the potential of other renewable energy sources. We will discuss this further. Of course, it should be noted that some greenhouse gases are produced during the installation, maintenance, and decommissioning of solar energy systems. Still, their amount is much less than the emission of harmful gases related to fossil fuels. An effective way to reduce carbon dioxide emissions at the same time as the growth of energy consumption will be a significant increase in the use of renewable energy, especially solar energy, for electricity generation, transportation, heating, cooling, and other applications.

The International Energy Agency has recently proposed different approaches for the global response to the risks of climate change. The plan is to reduce carbon dioxide emissions during energy conversion in 2050 to less than half their 2011 levels. The International Energy Agency predicts that emission reductions will be realized at a lower cost, and one of the scenarios considered is that non-economic factors limit the growth of nuclear power. The agency predicts global electricity demand will increase by 79 percent between 2011 and 2050. Wind, hydro, and solar power will account for 66 percent of global electricity generation in 2050, and solar power alone will provide 27 percent of the world's electricity. In the future, the increasing use of hydropower systems will likely be limited for environmental reasons, in which case solar energy should play a more significant role in providing global electricity so that less carbon dioxide is emitted. Today, there are environmental problems in developing hydropower facilities in many countries, and it is predicted that these problems will become more acute in the future.

In the following, we discuss three obstacles to the progress of solar systems: cost, scalability, and intermittency. Firstly, the cost of solar electricity has decreased drastically in recent years, and it can be expected to reduce even more. Now, in many countries, the use of solar energy for electricity generation is still more expensive than the use of fossil fuel power generation systems.

In addition, solar energy is artificially cost-disadvantageous since fossil energy users do not pay any damages for the emission of carbon dioxide and other gases. Applying such a scenario and comprehensive policies for users can significantly reduce the emission of harmful greenhouse gases. If the use of solar energy increases, the average cost of solar electricity will decrease because the cost of supplying this solar energy will decrease during sunny hours and especially during peak hours of solar radiation, when the most solar energy is produced. This process means that even where solar generation is more expensive than or equal to fossil energy generation, its cost must be reduced significantly to be cost-effective at higher levels of use.

Second, if solar energy becomes the primary source of electricity production by the middle of this century, solar technology, industry, and its supply chain will have to develop significantly. In the International Energy Agency scenario, solar electricity production will increase to more than 50 times 2013 by 2050. Of course, the development of some solar technologies has problems due to rare materials. Scaling up production to this magnitude will likely take a lot of work. Based on what has been observed, the limitations of materials like silicon are not a serious threat to the emerging solar energy industry.

Third, solar energy is intermittent and unstable; one of the characteristics of solar radiation is its unpredictability and intermittentness. This feature is one of the significant obstacles to using solar energy on a large scale and needs to be more reliable. It is essential to match the production and demand for power systems. In addition to the unpredictable production fluctuations, demand fluctuations are also entirely unexpected. Small-scale solar energy production does not pose a big problem. But in a power system heavily dependent on solar energy, the intermittency of the solar source makes the net load (the amount of electrical energy that the energy sources must supply) more variable and less predictable. On a smaller scale, according to the IEA scenario, most systems can overcome the intermittent nature of solar energy and the variability of production and demand loads by using more flexible fossil fuel generators. However, in most energy systems, especially if solar energy is used significantly, the development of large-scale energy storage technologies is needed.

By any standard, the solar energy source is enormous. Today, a tiny percentage of the world's land area is used for solar power plants. For example, in the United States of America, at least one-third of the country's area is used to produce corn, from which ethanol is produced. It provides only 7% of the country's energy for gasoline production. Whereas, if only 0.4% of the country's area is allocated for solar power plants and these power plants receive half of the annual solar radiation capacity, the entire country's energy will be supplied for one year. The area of use for solar power plants in America in 2021 is 0.027% (1034 square miles of 3.8 million square miles are dedicated to solar power plants).

On a global scale, solar resources are widely distributed and abundant. Wherever there are people, sunlight is available.

Figure 11 shows a map of the distribution of solar energy sources on the planet. Also, the histograms of land area, population, and average solar radiation are shown as a function of latitude and longitude. Notably, according to the map and images, no other resources, such as fossil fuel and suitable locations for wind or hydroelectric power generation, are distributed.

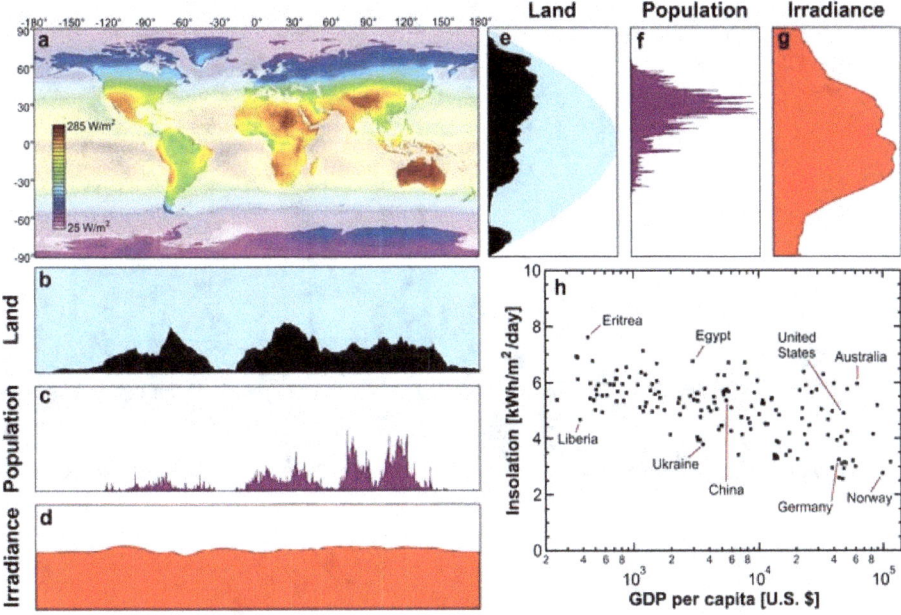

Figure 11: Distribution of solar energy sources in the world [9]

Figure 11 - a) shows the average intensity of solar radiation for different parts of the globe. Figure b) shows the amount of land area (black areas of land and blue areas show oceans), Figure C shows the population, and Figure D) shows the average solar radiation along the longitude. Figures e to g show these along latitude. Figure h shows the relationship between average radiation and GDP per capita [4] for countries worldwide in 2011. The range of changes in GDP is more than in the moderate intensity of solar radiation, and countries with low economic status have good access to solar energy. It can also be seen that the intensity of solar radiation is higher in the areas of the earth where the population is high. In the southwestern United States of America, Europe and South Asia, Australia, and Africa, almost all areas are suitable for solar power plants due to more solar radiation.

SOLAR ENERGY TECHNOLOGIES

The necessity of transition to sustainable and environmentally friendly energy sources has been revealed in recent years. One of the best ways to achieve these goals is to use solar energy differently. Technology and tools are needed to use solar energy. In general, there are five technologies for using solar energy. These five technologies are photovoltaic systems (conventional and thin), concentrated solar or thermal energy, solar heating and cooling, solar fuels, and passive solar design. This section explains the different types of solar energy technologies, and photovoltaic systems and concentrated solar energy (or solar thermal energy) are examined in more detail. The various solar energy applications for each of these technologies are mentioned in the following sections.

Figure 12 shows a picture of a photovoltaic power plant in China from a distance. Figure 13 shows the image of a concentrated solar power plant of the heliostat type.

Figure 12: A photovoltaic power plant in China [10]

Figure 13: Heliostat concentrated solar power plant (CSP or solar thermal power plant) [11]

A solar thermal power plant is usually used on a large scale with a production capacity of, for example, 100 megawatts. This type of power plant can store the sun's thermal energy for hours with low sunlight at night and use the stored energy whenever needed.

Photovoltaic systems can be installed on both large and small scales. The small scale of these systems is, for example, residential photovoltaic installations on the roofs of residential buildings, whose power generation capacity is usually less than 10 kW [12]. Alternatively, for example, a small photovoltaic panel that provides electrical energy for traffic signs in a remote area. The amount of electricity these photovoltaic systems produce depends on changes in the intensity of the sun's radiation. A photovoltaic system can use all the sun's radiation and produce electricity, but a concentrated solar energy system uses only direct radiation. As a result, the solar radiation receiver of a concentrated solar energy system is more sensitive to the presence of clouds, fog, and dust. However, its ability to store thermal energy makes its output more stable than a photovoltaic system. As a result, a photovoltaic power plant for cloudy areas is more economical than a solar thermal power plant because it requires fewer thermal energy storage devices.

Photovoltaic systems

Photovoltaic systems are the most well-known solar energy technology. A solar cell, a photovoltaic cell, directly converts sunlight into electricity. The word photovoltaic was mentioned for the first time around 1890 and comes from the Greek words photo, "phos," which means light, and "volt," which refers to electricity. The first modern solar cells were produced in 1954 and installed on a US space satellite in 1958.

Figure 14 shows the schematic of the electricity production components from the photovoltaic panel to its consumption. Figure 15 describes the schematic of implementing a photovoltaic system on the roof of a house. Photovoltaic panels convert a significant part of the sun's energy into electrical energy. These panels are designed to absorb sunlight effectively and convert it into low-voltage direct-current electricity that various household appliances and electrical devices can use. A critical component in the system is the battery to ensure a continuous and stable power supply. The excess electricity the photovoltaic panels produce during the sun's peak hours charges the battery. The battery acts as an energy reservoir and stores excess electricity for later use when sunlight is unavailable or can be sold to the national grid when energy demand increases. Another vital component is the hybrid inverter that performs multiple functions. The inverter, the system's heart, converts the DC electricity the photovoltaic panels produce into alternating current electricity. Alternating current (AC) is the standard form used for most home electrical appliances, and almost all electricity is transmitted as alternating current (AC) in power transmission and distribution systems. When the PV system produces more electricity than the household needs, the excess power can be fed into the grid. This integration with the grid, controlled by the hybrid inverter, allows homeowners to benefit from grid metering or feed-in tariff programs. For example, during peak consumption hours, when energy is expensive, sell this energy to the grid.

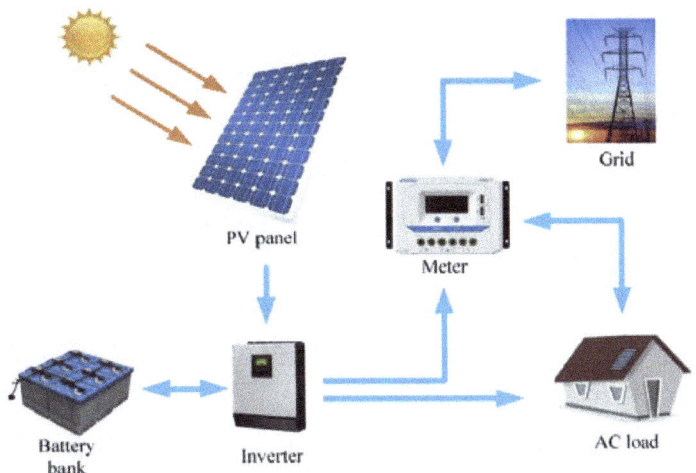

Figure 14: Schematic of electricity production components from a photovoltaic panel to consumption

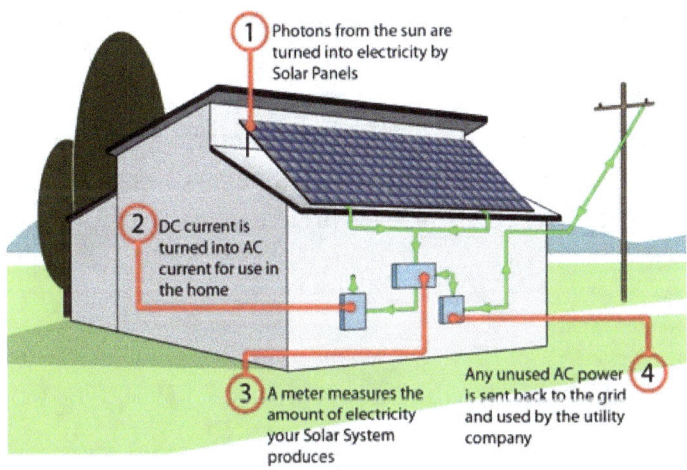

Figure 15: Schematic of the implementation of a photovoltaic system on a roof house

In summary, the sequence of electricity production in a photovoltaic system on the roof of a house or a large photovoltaic power plant includes

- converting solar energy into DC electricity through photovoltaic panels
- storing excess energy in batteries for future use
- converting the electricity to AC through a hybrid DC power inverter

This process allows homeowners to generate clean energy, earn money, contribute to environmental sustainability, and move toward a greener future.

A photovoltaic cell is usually small (about a few centimeters) and produces about 1 or 2 watts of direct current power. These cells are made of different semi-conducting materials, and often, their thickness is less than four human hairs. Some photovoltaic cells can convert artificial light into electricity. The structure of a solar array is shown in Figure 16. A solar array, which generates high power, consists of several modules. The module consists of several solar cells that can be connected in series and parallel. The efficiency of these modules is higher in the cold, i.e., in the winter, than in the summer, and over time, their efficiency decreases, like batteries. The efficiency of these modules varies based on the manufacturing technology, but it is 20% on average.

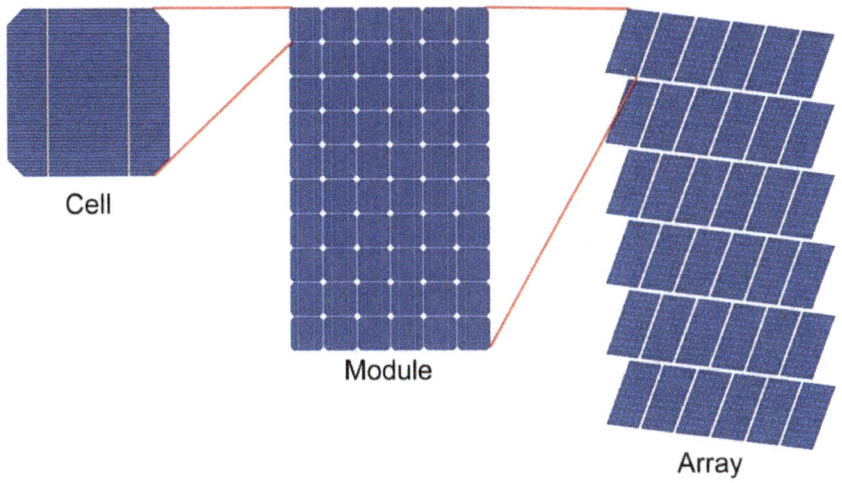

Cell

Module

Array

Figure 16: Cell, panel (module), and solar array [13]

Sunlight consists of photons or particles of solar energy. These photons contain different amounts of energy. When photons strike a photovoltaic cell, they may be reflected from the cell, pass through it, or be absorbed by the semiconductor material. Only absorbed photons provide energy to generate electricity. Electrons are ejected from the material's atoms when semiconductor materials absorb enough sunlight. The particular behavior of the material's surface during manufacturing makes the front surface of the cell receptive to displaced or free electrons so that the electrons migrate naturally to the cell surface. The materials utilized were deliberately fabricated to exhibit precise behavior such that electrons are intrinsically transported to the exterior surface of the cell. The movement of electrons toward the front surface of the solar photovoltaic cell causes an imbalance in the electrical charge between the front and back surfaces of the cell. This imbalance creates a voltage potential like a battery's negative and positive terminals. The electrical conductors on the cell absorb the electrons. Electricity flows through the circuit when the conductors in an electrical circuit are connected to a battery or, for example, a light bulb. Figure 17 shows the structure and function of a photovoltaic cell.

Inside a photovoltaic cell

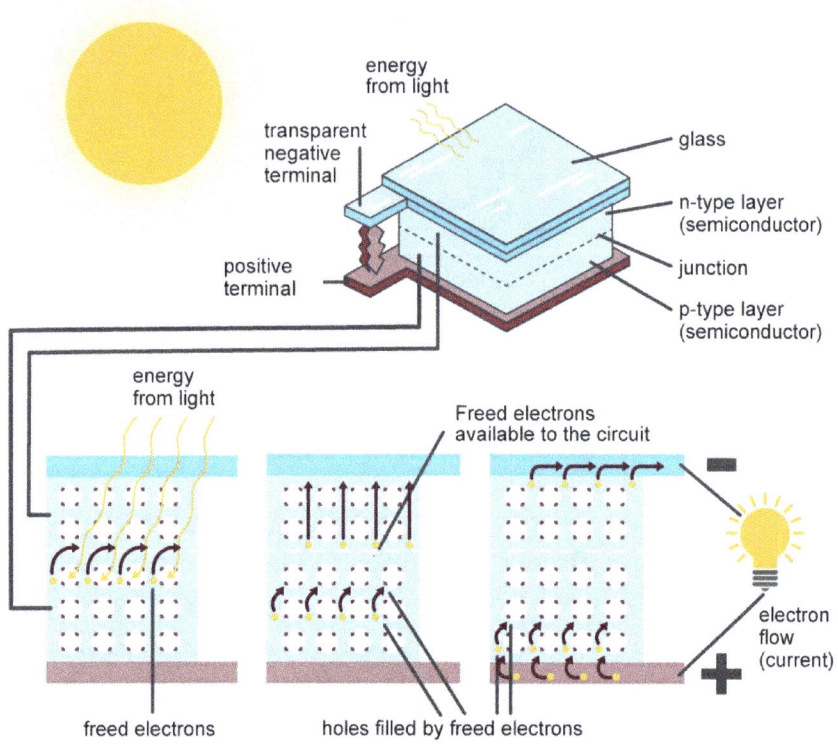

Source: U.S. Energy Information Administration

Figure 17: Structure and function inside a photovoltaic cell

The production of photovoltaic cells, in turn, has different technologies; as a result, each has its specific applications. Some of the most essential technologies for making these cells are summarized as follows:

Mono-Si (Mono-Si): This technology uses high-grade silicon with a mono-crystalline structure, resulting in good efficiency and performance. They are commonly used in residential and commercial applications.

Polycrystalline Silicon (Poly-Si): Polycrystalline cells are made of multiple silicon crystals, which reduces manufacturing costs.

They are less efficient than mono-crystalline silicon but are widely used because they are cheap.

Thin-film solar cells: Thin-film solar cells are made by placing a thin layer of semiconductor material on a substrate such as glass or metal. This technology offers flexibility, lightweight design, and lower manufacturing costs. Materials such as amorphous silicon or cadmium telluride are used for construction.

Figure 18: A flexible thin film photovoltaic cell [14]

Concentrated Photovoltaic Systems: These systems use lenses or mirrors to focus sunlight onto small, high-efficiency solar cells. These systems can achieve higher conversion efficiency by concentrating sunlight than conventional photovoltaic systems.

Double-sided solar panels: Double-sided panels are designed to absorb sunlight from both sides and increase energy production. They can absorb direct sunlight from the front and reflect sunlight from the back. Double-sided panels can be used in various applications, such as solar boats and rooftop installations.

Perovskite solar cells: Perovskite solar cells are thin film solar cells that use a compound with a perovskite structure as a light-harvesting layer. They have shown rapid improvements in productivity and offer a potentially cost-effective and highly efficient solution. However, commercialization is still ongoing, and more research is needed.

Solar Tracking Systems: Solar tracking systems adjust the position of solar panels to follow the sun's path throughout the day and maximize energy production. Tracking systems can be uniaxial (following the sun's movement from east to west) or biaxial (following the movement east-west and up-down).

Photovoltaic combined with building: These types refer to integrating solar panels in building materials such as roofs, windows, and facades. This technology enables the seamless integration of solar power generation into building design, offering functional and aesthetic benefits.

Figure 19: An integration of building materials with photovoltaic cells [15]

Photovoltaic installation cost is usually divided into two parts: the cost of the solar module and the so-called system balance costs, which include the costs of converters, hardware installation racking, design work, installation costs, marketing as well as monitoring costs, and Financing is different. Photovoltaic technology choices affect module costs and system balance. After decades of development, supported by significant research and development investments, today's leading photovoltaic solar technology, wafer-based crystalline silicon[5], has matured and developed in terms of technology, and there is a large-scale production capacity of crystalline silicon modules. However, current crystalline silicon technologies also have inherent technical limitations, the most important of which include high processing complexity and low inherent light absorption. High intrinsic light absorption requires a thick silicon wafer. The strength and weight of the glass-encased crystalline silicon modules help balance the cost of the system.

Companies manufacturing crystalline-silicon modules and their component cells and materials are motivated to pursue opportunities to make this technology more competitive by improving efficiency and reducing manufacturing and material costs. Therefore, government support for research and development in the current crystalline-silicon technology is unjustified.

Limitations in the availability of crystalline silicon have led to research into thin-film photovoltaic alternatives. Commercial thin-film photovoltaic technologies, mainly cadmium telluride and indium gallium diselide solar cells, account for approximately 10% of today's photovoltaic market and are already economically competitive with silicon. Some commercial thin-film technologies depend on rare elements, making it challenging to generate practical terawatt-scale electricity. For example, the abundance of tellurium in the Earth's crust is estimated to be only one-fourth that of gold. Several emerging thin-film technologies currently under investigation employ new materials and structures and have the potential to provide better performance with lower manufacturing complexity and cost [16]. A number of these technologies use materials that are abundant on Earth. Other features of some of the new thin-film technologies include low weight and compatibility with mounting in flexible forms, resulting in lower system balance costs and module costs.

Concentrated solar power (solar thermal power plant)

One promising solution gaining increasing appeal is concentrated solar energy, also known as solar thermal energy. As a renewable energy technology, concentrated solar energy offers tremendous potential to meet our energy needs while minimizing environmental impact. As the name of this technology suggests, the goal is to collect and concentrate solar rays so that valuable things can be done with them.

This technology (except for its dished version described below) is usually used for large-scale, urban power generation. Unlike traditional photovoltaic cells that directly convert sunlight into electricity, the sun's heat is absorbed and stored in concentrated solar power. This heat can be absorbed by a fluid such as water, industrial oils, air, or solid materials such as rock. The absorbed heat energy can start an electricity generation cycle, a power plant that produces electricity from the sun's heat (concentrated solar power plant). Various methods, including coal burning, natural gas, nuclear reaction, and others, provide the heat needed in electricity generation plants. As mentioned, a solar power plant can also provide this heat. The selection of the names of the types of these power plants is based only on the way they supply thermal energy. Otherwise, continuing the electricity production process is common to all of them.

Advances in concentrated solar energy technologies enable us to generate clean electricity and provide opportunities to incorporate thermal energy storage. Excess heat generated during peak solar radiation can be stored in phase change materials by integrating thermal energy storage systems. This stored energy can then be released during low sunlight or at night, ensuring constant electricity production. Several technologies are associated with concentrated solar energy, each with unique functions and benefits. The device that collects solar thermal energy is called a collector, which has different types. Some of these technologies, each of which uses its collectors, are summarized as follows:

Parabolic solar collector systems: This system consists of large, curved mirrors that focus sunlight onto a long receiver tube located along the mirrors' focal line. A heat transfer fluid flows inside the receiver tube, absorbing concentrated sunlight to produce water vapor in later stages. The power plant's turbine moves with the help of this steam at high temperatures and pressures to generate electricity. This technology is widely used in large-scale and municipal power plants due to its proven efficiency.

Figure 20: A solar parabolic collector in California, USA [17]

Solar power towers: These towers use a set of mirrors called heliostats to focus sunlight on a central receiver at the top of the tower. Focused sunlight heats a fluid in the receiver. This thermal energy can be used to produce electricity, which is almost the same process as producing electricity with other power plants. This process is explained in the previous sections. The advantage of this technology lies in its ability to achieve higher temperatures and, thus, higher energy conversion efficiencies.

Figure 21: A heliostat solar power tower [17]

Linear Fresnel reflectors: Linear Fresnel reflectors use flat mirrors to focus sunlight onto a receiver tube. These mirrors are arranged in long, parallel rows and follow the sun's movement throughout the day. Similar to the previous explanation, the absorbed heat is used to generate electricity. Linear Fresnel reflectors are cheaper due to their straightforward construction.

Figure 22: A concentrated solar power plant of linear Fresnel type [18]

Stirling dish systems: Stirling dish systems consist of a parabolic dish that reflects sunlight to its focus. A heat receiver is located in the center of this dish and contains a Stirling motor. The engine uses the absorbed heat to drive the piston. Thermal energy is converted into mechanical energy and electricity through a generator. These systems are particularly suitable for small-scale applications, such as off-grid power generation or remote locations.

Figure 23: A concentrated solar power system of the dish type [19]

Figure 24: The process of concentrating sunlight in a concentrated solar power system at its parabolic focus, dish type [20]

Concentrated solar power systems can be deployed on a large scale without facing problems in the supply of raw materials. Also, the ability to include thermal energy storage in these systems means that concentrated solar power can be a dispatchable power source. Dispatching is used in large power plants and power networks and controls all parts of these power plants, including production, transmission, and distribution. When electricity is needed, and the electricity system faces a shortage, the required electricity can be supplied from the stored heat. The best ways to improve the economic costs of concentrated solar power are higher operating temperatures and more efficient solar energy collection methods.

Therefore, research and development expenditures in concentrated solar power technology should focus on improvements in system design and solar energy capture. For example, work on systems such as solar towers, in materials science operating at higher temperatures, and developing systems for energy harvesting can be very useful.

Concentrated solar power holds promise as a sustainable energy solution and offers numerous environmental benefits. This clean, abundant, and renewable resource can reduce our dependence on fossil fuels, reduce greenhouse gas emissions, and mitigate climate change. Ongoing research and development efforts continuously enhance the efficiency, affordability, and scalability of concentrated solar power systems, thereby solidifying their pivotal role as a significant contributor to the future of clean energy.

APPLICATIONS OF SOLAR ENERGY

For centuries, humans have used this abundant and renewable energy source for various applications. With technological advancements, solar energy has emerged as an efficient and sustainable solution to meet ever-increasing energy needs while mitigating environmental concerns. Solar energy has found many new and increasing applications in different sectors. Some of these applications are briefly categorized as follows:

1- Electricity production: The applications of this field are very diverse because today, most devices work with electrical energy.

I) Photovoltaic systems: Photovoltaic systems, also known as solar energy systems, have wide applications in different sectors. Some of these main applications are a solar panel farm, used in remote locations, self-sufficient power systems, space applications, building needs, and military and transportation applications. These uses and some other examples are explained below:

a) Residential: Photovoltaic systems can be used in residential buildings to generate electricity for domestic needs. They can provide electrical appliances, lighting, heating, and cooling, reducing dependence on the national electricity grid and the cost of energy bills. Photovoltaic panel solar farms can provide power on a practical scale, from tens of megawatts to more than one Gigawatt(GW). These large systems feed electricity into municipal or regional grids using fixed panels or solar trackers. Photovoltaic materials can also be integrated into a building's structure as windows, roof tiles, or cladding to generate power and serve as a utility structure. Also, canopies and parking structures can be covered with photovoltaics to provide shade and power.

b) Commercial and Industrial: Photovoltaic systems are widely used to offset energy consumption in commercial and industrial environments. Appropriate arrays of solar panels can be installed on rooftops, parking lots, or open spaces to generate clean energy for offices, factories, warehouses, and shopping centers.

c) Off-grid systems: Photovoltaic systems are valuable in remote areas without access to the power grid. Together with energy storage solutions such as batteries, these systems provide an independent power supply for lighting, communication devices, water pumps, military applications, remote bases, and other essential appliances.

d) Grid-connected systems: Photovoltaic systems can be combined with the power grid, which allows excess power to be fed back into the grid. This process will enable users to earn credit through net metering or feed-in tariffs while ensuring a stable power supply during cloud periods or high-demand conditions. Net metering is a system where solar panels or other renewable energy generators are connected to a public power grid, and excess power is transferred to the grid, allowing customers to offset the cost of electricity taken from the grid.

e) Agriculture: Photovoltaic systems have applications in agriculture, such as irrigation systems, water pumping, and livestock operations. Solar fences, warehouse or stable lighting, and automatic feeding systems can also be implemented using photovoltaics.

f) Transportation: Solar panels integrated into electric vehicles, including cars, buses, airplanes, and bicycles, can help charge the vehicle's battery and increase its range. Solar charging stations along highways or parking lots can also support the adoption of electric vehicles.

g) Water purification: Photovoltaic systems can be combined with water purification technologies such as reverse osmosis or UV disinfection. Solar energy water purification systems are beneficial in areas with limited access to drinking water.

h) Portable applications: Solar chargers, portable solar panels, and solar backpacks are popular and practical for charging mobile devices, laptops, and camping equipment. These programs can provide stable power in external environments.

i) Natural disaster relief: In emergencies or natural disasters, photovoltaic systems provide temporary electricity to critical infrastructure such as hospitals, emergency response units, and relief camps. In these cases, regular power sources are usually interrupted.

j) Space applications: Photovoltaic systems are widely used in space missions to supply energy to satellites, spacecraft, and the International Space Station. Solar panels absorb sunlight in space and convert it into electricity to support various operations. These panels are widely used in various space missions and satellites to provide energy for communication systems, scientific instruments, propulsion systems, and other onboard equipment. Due to the abundance of sunlight available in space, photovoltaic panels provide a sustainable and renewable energy source that reduces dependence on other limited resources and minimizes the environmental impact of space exploration. These panels' compact and lightweight nature makes them ideal for space applications where weight and size constraints are critical factors. In general, photovoltaic panels are vital in providing energy for spaceflights, enabling long-duration flights, facilitating research efforts, and expanding our understanding of the universe beyond Earth's borders.

k) Military applications: Photovoltaic panels, especially their thin and light type, have various military applications, including providing the required power for soldiers by charging batteries and starting essential equipment. Soldiers are stationed at remote military installations, providing self-sufficient power generation for lighting, communications systems, and surveillance equipment. Photovoltaic panels are used for power without driver systems, including drones and ground vehicles, extending their mission duration without refueling or replacing batteries. These panels are portable power sources for charging handheld devices and radio batteries while facilitating field communications by powering radio repeaters, satellite terminals, and mobile command centers. Moreover, their implementation contributes to enhanced energy independence and decreased reliance on external power grids. These panels provide portable power sources for battery charging of handheld devices and radios and support field communications by powering radio repeaters, satellite terminals, and mobile command centers. In addition, their deployment increases energy independence and reduces dependence on external power grids.

II) Thin film photovoltaic cells: Thin film solar cells are a type of photovoltaic cell that uses thin layers of semiconductor materials such as amorphous silicon to convert sunlight into electricity. Due to their flexibility and thinness, they have many applications.

III) Concentrated solar energy or solar thermal energy systems: These systems use mirrors or lenses to focus sunlight on the solar energy receiver, which converts it into heat. The heat is then generated through steam turbines or other heat engines. These systems usually produce electricity on a rural and urban scale.

2- Solar heating: Solar heating is an environmentally friendly way to use the sun's free energy to heat homes, buildings, and industrial devices. The way these systems absorb the sun's heat is similar to the absorption of heat in concentrated solar power plants, which was explained. Solar air heating systems collect the sun's radiation and use it to heat air or any other fluid, such as water, that can be used for space heating or industrial processes. These systems are often used in buildings, greenhouses, drying applications, solar desalination, and solar cooking.

3- Solar cooling: Solar cooling systems use solar energy for cooling and air conditioning. These systems are designed to reduce dependence on the national electricity grid and reduce carbon emissions associated with conventional cooling technologies. After going through processes, these systems absorb the sun's heat energy and fabulous homes, offices, and industrial facilities.

4- Solar fuels: Solar fuels are a promising solution in the search for clean and sustainable energy sources. Solar fuels, such as photovoltaic or photoelectrochemical systems, are produced through direct or indirect solar energy conversion. This technology uses solar energy to perform chemical reactions to produce usable fuel. These fuels that use solar energy have several advantages. First, solar fuels are a carbon-free alternative to conventional fossil fuels and help reduce the adverse effects of greenhouse gas emissions and climate change. Unlike fossil fuels that release carbon dioxide when burned, solar fuels only produce water and oxygen, so they are environmentally friendly and compatible with the circular economy[6]. Another significant advantage of solar fuels is their potential for energy independence. By capturing and storing solar energy to fuel, we can reduce our dependence on limited and often geopolitically sensitive fossil fuel sources. In addition, solar fuels have considerable versatility in terms of their applications. Hydrogen is one of the most popular solar fuels that can be easily stored and used in various industrial processes and power generation systems. Other solar fuels include methanol, ammonia, synthetic natural gas, carbon-based fuels such as gasoline, diesel, jet fuel, alcohol, and sulfur-based fuels such as hydrogen sulfide. Solar fuels such as hydrogen can also be integrated into existing energy systems, facilitating the transition to a sustainable and decentralized energy grid.

The development of solar fuel technologies has seen significant progress in recent years. Researchers are exploring different approaches, such as artificial photosynthesis, which mimics the natural process of photosynthesis to convert sunlight, water, and carbon dioxide into fuel. Materials engineering is vital in designing and optimizing efficient solar fuel devices, including photovoltaics, catalysts, and photoelectrodes. While solar fuels have great potential, their large-scale deployment still faces challenges. Research and development efforts are focused on improving solar fuel technologies' efficiency, durability, and affordability. In addition, building a solid infrastructure for producing, storing, and distributing solar fuels is essential to ensure widespread adoption. Overall, solar fuels offer an exciting path to a cleaner and more sustainable energy future. These fuels use the sun's power and are compatible with the environment.

They provide zero and carbon neutrality to meet our energy needs. Continued advances in solar fuel technologies promise to transform our energy landscape and empower us to create a more resilient and sustainable world for future generations.

5- Passive solar design: Passive solar design is an architectural technique to maximize the use of sunlight for heating, cooling, and lighting of buildings. The goal is not to use mechanical systems, which need to be purchased and used, which require cost and energy. Features such as large south-facing windows, thermal mass, and awnings help optimize energy efficiency. Critical elements of passive solar design include orientation, glazing and insulation, thermal mass, shading, ventilation, and lighting. Place the windows of the building in the direction of the sun to receive maximum sunlight. Glass and insulation, for example, use low-consumption windows and insulation to prevent heat loss in colder seasons and maintain cool air inside buildings in hot months. Thermal mass, for example, uses materials with high heat storage capacity, such as concrete or stone, to absorb and slowly release heat to stabilize the temperature inside the building. They were shading, such as implementing projections, canopies, or vegetation to control the amount of direct sunlight that enters the building in the hot seasons and prevent overheating. Ventilation means ensuring adequate airflow by placing windows, vents, and thermal chimneys to facilitate natural cooling and reduce dependence on mechanical systems. Lighting is to maximize the penetration of natural light through windows, skylights, and skylights that minimize the need for artificial light. Through the utilization of passive techniques, buildings have the potential to mitigate energy consumption while simultaneously enhancing occupant comfort effectively. This process makes notable contributions towards attaining sustainability objectives and energy efficiency goals.

In the following, we will further investigate solar energy from different aspects necessary to understand this type of energy and make better decisions.

Photovoltaic installation

The recent significant growth in the use of photovoltaic systems is due to the reduction of at least 50-70% of reported photovoltaic prices (in the United States of America without federal subsidies) per installed peak watt [21]. Almost all of these improvements result from reduced prices of modules and inverters. In addition, the structure of the solar energy market, especially at the residential level, is changing with the evolution of new business models, the introduction of new financing mechanisms, and the impending reduction of subsidies in various countries worldwide.

Solar Economy

The economic competitiveness of solar electricity compared to other electricity generation technologies depends on its cost and output value in the electricity market. A measure usually used to compare different sources of electricity is the levelized cost of electricity [22]. However, the levelized cost of electricity needs to be revised to assess the competitiveness of photovoltaics or to compare photovoltaics with concentrated solar power or conventional generation sources because the amount per kilowatt-hour of photovoltaic generation depends on many characteristics of the regional electricity market, including the level of photovoltaic penetration. For example, the more photovoltaic capacity in a given market, the less valuable it is to increase photovoltaic generation.

Solar utility scale

Nevertheless, estimates of the levelized cost of electricity are helpful because they provide a rough idea of the competitive position of solar energy at its current low penetration level in the electricity supply mix. In the economic evaluation of utility-scale solar generation, an appropriate point of comparison is with other utility-scale generation technologies, such as natural gas combined cycle power plants [23]. For example, without considering the cost of carbon dioxide emissions and federal subsidies in the United States, current industrial-scale photovoltaic electricity has a higher levelized cost than combined cycle natural gas generation in most regions of the United States, including relatively sunny areas. It is like Southern California.

Designing concentrated solar power plants with thermal energy storage reduces the levelized cost of electricity. It allows these plants to produce electricity when electricity is most valuable, making them more competitive with other sources of electricity generation. However, utility-scale photovoltaic generation is about 25% cheaper than concentrated solar power.

Residential solar

Suppose solar generation is valued for its share at the system or wholesale level, assuming that solar penetration into the grid does not cause a net increase in distribution costs. In that case, residential PV generation is, on average, about 70% more expensive than large-scale PV generation. - It is beneficial. Now, even in California, and even with 100 percent effective federal subsidies, residential PV is not competitive with natural gas combined cycle generation on a levelized cost of electricity basis.

Lowering the balance of system costs to more typical levels of PV installations in Germany brings residential PV closer to a competitive position. However, residential PV is still more expensive than photovoltaic or city-scale natural gas combined cycle power plants.

In most electricity distribution systems, grid-connected residential photovoltaic systems offset generation under net metering. In this flow, the owner of the residential PV installation pays the residential, retail rate for electricity purchased from the local distribution company and is compensated at the same rate for any excess PV output fed back into the grid.

In this situation, a commonly used investment criterion is grid parity, achieved when it is equally attractive and cost-effective to use a rooftop photovoltaic system to meet a portion of a residential customer's electricity needs.

Combining electricity produced from photovoltaic solar energy with existing power plants
Distributed use of solar energy

The use of photovoltaic electricity in a distributed manner (for example, residential buildings in different parts of the city are equipped with photovoltaic electricity generation systems) imposes costs on the system in two ways. In general, electrical energy losses in the transmission network are initially reduced by increasing the use of distributed photovoltaic electricity. However, when using more distributed photovoltaic energy and increasing a significant share of the total production, it causes the distribution costs to increase. As a result, local rates will also increase because additional investments are needed to maintain the quality of electricity. After all, the currently used networks are not designed to manage and organize the return of electricity from customers to the grid [24]. Electricity storage is a costly alternative to grid reinforcement or distributed photovoltaic power surge management upgrades.

In an accurate, efficient, and fair distribution system, each customer pays a share of the costs of the distribution network, which reflects his responsibility for creating these costs. Instead, most U.S. utilities categorize distribution networks, electricity, and other fees and then charge a flat rate per kilowatt-hour that covers all those costs.

This change in costs and subsidies for residential solar generators has already led to political conflicts in some regions and states of the United States of America. These conflicts are expected to intensify with the increase in the use of solar energy in residential houses. Rapid and long-term growth in distributed solar power generation will likely require consideration and developing pricing systems that lead to efficient and fair investment in the grid.

Wholesale and macro markets
for solar energy

When combined with thermal energy storage, concentrated solar power generation can be used and dispatched like thermal or nuclear power generation in electricity markets.

However, coordination and dispatching challenges arise when PV power has a significant share of the wholesale electricity market. In nearly two-thirds of the United States and many other countries, generators auction their electricity generation to competitive wholesale markets. Photovoltaic units offer an auction with their final cost of production, which is zero, but on the other hand, they earn the final price and profit of the system every hour.

In wholesale electricity markets, it replaces the systems of conventional electricity producers (e.g., fossil fuels) with the highest fluctuating costs. This process reduces variable production costs and thus lowers market prices. Moreover, since older, conventional systems depend on fossil fuels, this also results in lower carbon dioxide emissions.

In the future, the combination of non-solar energy generation with large volumes of solar energy at maximum capacity will likely be adjusted more flexibly. Therefore, the economic value of photovoltaic output can increase. Also, netload peaks can be reduced, and cycle requirements on thermal power generators can be reduced by coordinating solar power generation with hydroelectric power, pumped storage, other available types of energy storage, and demand management techniques. Because of the potential importance of energy storage in facilitating more significant use of solar energy, large-scale storage technologies are attractive subjects for R&D spending.

It is not cost-effective to use photovoltaic energy at low volumes; even if it becomes cost-effective to produce this energy at low penetration levels, the revenue per kilowatt of installed capacity decreases as solar penetration increases until a break-even point is reached, beyond which the investment Most of the solar photovoltaic energy will be useless. Therefore, a significant cost reduction is needed to make the photovoltaic system competitive in a vast volume of use, which is foreseen in many plans to reduce carbon dioxide production.

This solar penetration is reduced in systems that require long-term energy storage, such as hydropower plants with large reservoirs. As opportunities for new hydroelectric power generation or pumped storage are limited, breaking the barrier to high solar power usage further highlights the importance of development. Economic technologies of multi-hour energy storage are also an excellent approach to using cost-effective photovoltaic energy on a large scale.

Establishing and using the technology of solar energy

Reasons cited to support the allocation of subsidies for current solar technology usually include short-term greenhouse gas emission reductions and job creation. However, the main goal should be solar power generation's large-scale, long-term growth to significantly reduce future greenhouse gas emissions and meet the growing global energy demand. In the second stage, the main goal is to achieve these two goals with the most effective use of budget public capital and private resources. The least expensive way to promote solar energy is through one of several price-based policies that reward solar energy output according to its value to the electricity supply system.

Subsidies for solar technologies are far more effective per taxpayer dollar spent if they reward production than investment. This change rectifies certain inefficiencies, specifically the disparate subsidy allocation between a kilowatt-hour of electricity generated by a residential photovoltaic system and a kilowatt-hour generated by an adjacent city-scale power plant.

TRANSMISSION, STORAGE, AND DISTRIBUTION OF ENERGY

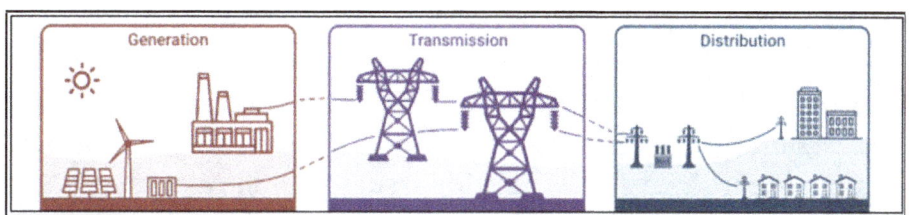

Figure 25: Generation, transmission, and distribution of energy

As the energy demand continues to grow, it is necessary to discover efficient ways to extract, transmit, store, and distribute energy. The energy supply process includes energy extraction, storage, transfer, and explanation. These are essential in ensuring an efficient and reliable energy supply to meet global demand. A complex network is critical to facilitate the seamless energy transfer from production to consumers, enabling efficient use and utilization of different energy sources.

Energy distribution is an essential aspect of the modern energy landscape, enabling the widespread access and consumption of different types of energy. From electricity to natural gas grids, district heating systems, and the integration of renewable energy sources, efficient distribution systems ensure a reliable and sustainable energy supply. By continuously improving and innovating these systems, we can optimize energy use, reduce environmental impacts, and bring the benefits of energy to all humans.

It is noteworthy that energy transfer refers to the movement of energy from one system or body to another system or body, which is transferred in the form of heat due to temperature differences. These methods are conduction, convection, and radiation, as explained in the previous parts of this chapter. The matter is needed for energy transfer by conduction and convection, but the matter is not needed for the radiation method. For example, solar energy passes through millions of kilometers without a vacuum and enters the earth through radiation without needing matter. Energy transmission is another concept discussed in this section [27].

In the previous parts, the way of generation, extraction, and conversion of energies was explained. In this section, we will examine more energy distribution issues, including transmission and energy storage, in a categorized manner.

Energy Transmission

It refers to the movement of energy over longer distances. For example, crude oil is transported between cities and countries through pipelines. This transportation enables energy distribution from the point of production to the end of consumption. It includes different electrical, thermal, mechanical, and chemical energy types. Whether through power cables, pipelines, or transportation systems, energy transmission ensures that energy is transported to power plants, industrial facilities, residential areas, and beyond. There are different methods for energy transfer, each of which has its characteristics, advantages, and applications [28].

Electric energy transfer:

Electricity is the most common form of energy transfer worldwide. This method transmits energy through power lines and cables and is accessible for lighting, heating, and powering various devices. Power grids enable efficient transmission and distribution of electricity over large distances. For example, high-voltage transmission lines carry electricity generated in power plants to substations and then distribute it to residential and commercial areas.

Advantages: Efficiency over long distances, manageable control and regulation of electric current, and easy conversion into other forms of energy, such as mechanical and thermal energy, through motors.

Disadvantages: Power losses occur over very long distances due to resistance; proper infrastructure maintenance is necessary to avoid blackouts and interruptions in energy transmission.

Pipeline transfer:

For oil, natural gas, and other hydrocarbons, pipelines serve as the primary means of transportation. This method is usually used for long-distance shipping. These networks transfer energy resources from extraction points to refineries or distribution centers. The Trans-Alaska Pipeline System is a pipeline that transports crude oil from the North Slope of Alaska in the United States of America to the coastal areas of Valdez, which is 1,287 km long. Natural gas is an efficient energy source for heating, cooking, power generation, and efficient distribution systems. Natural gas pipelines form an extensive network that connects natural gas producers to distribution centers and end consumers. These pipelines carry gas under high pressure, allowing it to flow long distances, often crossing international borders.

Advantages: Efficient and cost-effective for transporting large quantities of liquids, reducing the risk of spills and accidents compared to other methods via truck. Disadvantages: having a relative limitation to certain materials that are difficult to transport through pipes. Initial installation and maintenance are expensive and time-consuming.

Wireless energy transfer:

Wireless or remote energy transfer technologies are emerging as a potentially world-changing change. Their purpose is to transmit energy without physical connections, electromagnetic fields, or resonance induction. Nikola Tesla's pioneering work in wireless power transmission laid the foundation for future innovations in this field. These technologies include radio waves, inductive charging, microwave power transmission, and solar energy satellites.

Radio waves transmit signals to various electronic devices such as radios, televisions, and wireless Internet connections. Inductive charging technology is commonly used in wireless charging pads for smartphones and other devices. It uses electromagnetic fields to transfer energy between two coils, one in the charging pad and the other in the device. Microwave power transfer involves using microwaves to transfer energy from a power source to a device.

This method has been used experimentally to transfer energy over long distances and can be used for future wireless charging of electric vehicles. *Solar energy satellites* are equipped with large solar panels that can generate electricity from the sun. This energy is converted into microwaves and sent to Earth for use, but this has yet to be done.

Energy storage

As mentioned, some energy sources are intermittent. Energy storage is essential to solve this problem and ensure continuous power supply. Here are some examples of energy storage technologies [29]:

Battery storage: Batteries store energy in chemical form and convert it into electricity when needed. They are used in portable devices, electric vehicles, and grid-scale storage systems. It should be noted that lithium-ion batteries are widely used due to their energy density and high efficiency.

Pumped water storage: One of the most common storage methods is pumping water to high altitudes when we do not need energy. When the energy is required, it can be converted into electricity as a stored gravitational potential energy and supply periods of high demand for electricity, such as noon. This technology provides massive storage capacity and helps stabilize the power grid.

Hydrogen storage: Hydrogen can be used as an energy carrier and stored in underground tanks. It can be produced through electrolysis of water using excess electricity and later used to generate electricity through fuel cells or combustion.

Thermal storage: Thermal energy can be stored and later converted into electricity or used as a direct heat source. For example, these systems use molten salt to store heat from solar thermal power plants or phase change materials to store and release heat for heating systems.

Superconducting Magnetic Energy Storage: These systems store energy in the magnetic field created by a superconducting coil. The stored energy can be released if needed by converting magnetic energy into electrical energy.

Integration of renewable energies:

With the expansion of renewable energy sources such as solar and wind energy, integrating these intermittent energy sources helps provide a uniform energy supply to the grid. Advanced energy distribution systems enable efficient management and distribution of renewable energy by balancing intermittent supply and demand. For example, intelligent grids use real-time data and digital technologies to monitor and manage the flow of electricity across the distribution network, ensuring that additional renewable energy is incorporated into the grid when available and fills energy gaps.

Conclusion:

Efficient energy transmission, storage, and distribution systems are critical components of a sustainable energy framework. Understanding and developing these technologies can optimize energy consumption, reduce losses, and build a cleaner and more reliable energy infrastructure. Embracing these developments will lead us to a future where energy needs are met, and planet Earth is preserved for future generations.

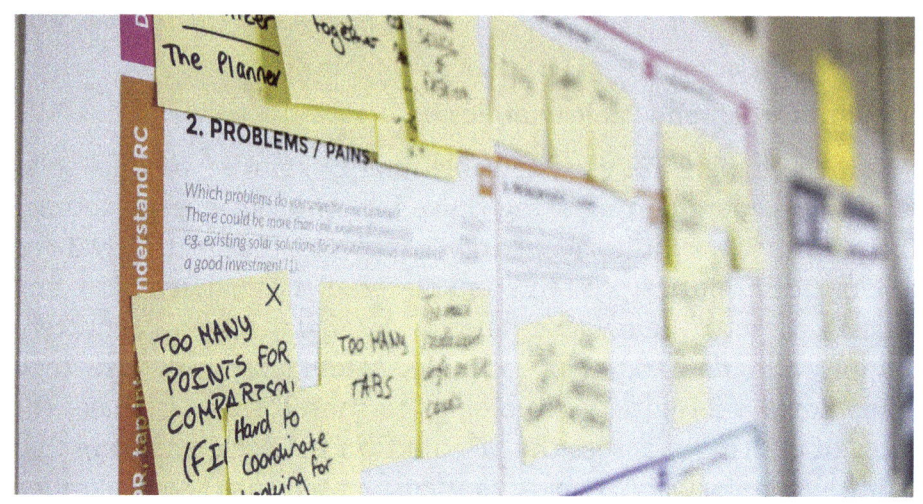

A FINAL APPROACH

In response to the pressing issue of global warming, solar energy emerges as a promising solution with significant long-term potential to fulfill human energy requirements, effectively mitigating greenhouse gas emissions and contributing to environmental preservation. Solar energy has emerged as a dependable and rapidly expanding means of electricity generation in certain regions worldwide. Governmental policies have facilitated its progress at the federal, state, and local levels in the United States, with substantial support from governments in Europe, China, and various other locations. The solar industry is developing and expanding with much more significant growth than other energy technologies. As a result, it can play a vital role in the global energy system by the middle of the century. As mentioned, the International Energy Agency has predicted that the growth rate of solar energy is so high that the share of the world's energy supply through solar energy will surpass all other sources by 2027 and reach 22.2%. However, this resource is only available in the same plentiful and affordable way for some parts of the world.

Although the cost of using solar energy has decreased significantly in recent years, and its market penetration and use have grown, its large-scale use is expected to increase in the coming decades. The solar industry's ability to overcome several obstacles depends on it. These barriers can include cost, intermittency, availability of technology and raw materials, and successful integration of the large-scale system into existing electrical systems. In general, government policies play an essential role in overcoming these challenges. However, research and development on solar energy technologies have shown that all obstacles are being removed. The costs of these systems are continuously decreasing. Suitable technologies for intermittent solar energy have been adequately developed. For the problem of intermittency, systems such as generators based on fossil fuels and hydropower plants can be combined with solar energy systems so that whenever the solar energy decreases, these systems compensate for the energy. Investigations so far show that raw materials are relatively easy to obtain and that large-scale development and integration are possible. Renewable energies become completely affordable and prioritized, especially if a penalty for carbon dioxide production and environmental degradation of fossil fuels is considered. Otherwise, the affordability of this energy and other Renewable energies are facing challenges in some countries. Among renewable energies, solar energy is more abundant and available than different energies in general. For example, problems related to dam construction capacity limitations, political tensions, and water shortages between countries have jumped to start hydroelectric power plants in recent years.

The rapid growth of solar energy and the world's transition to green energy largely depend on policymakers deciding to price the carbon emission penalty and implementing the right policies. The risks and challenges posed by global climate change and warming, and the potential for solar energy to play a significant role in managing these risks and challenges, may not be reason enough to modify and sustain efforts by law enforcement and the private sector to support solar energy technology.

REFERENCES

[1]   Renewables 2020 – Analysis - IEA n.d. https://www.iea.org/reports/renewables-2020 (accessed June 26, 2023).

[2]   World Final Energy - World Energy Data n.d. https://www.worldenergydata.org/world-final-energy/ (accessed June 26, 2023).

[3]   The sun as you've never seen it: European probe snaps closest-ever photo of our star | Space n.d. https://www.space.com/closest-ever-sun-photo-solar-orbiter (accessed August 3, 2023).

[4]   Ritchie H, Roser M, Rosado P. Energy statistics data. Our World in Data 2022.

[5]   Nasrabadi AM, Malaei O, Moghimi M, Sadeghi S, Hosseinalipour SM. Deep learning optimization of a combined CCHP and greenhouse for CO2 capturing; case study of Tehran. Energy Conversion and Management 2022;267:115946. https://doi.org/10.1016/j.enconman.2022.115946.

[6]   Share of cumulative power capacity by technology, 2010-2027 – Charts – Data & Statistics - IEA n.d. https://www.iea.org/data-and-statistics/charts/share-of-cumulative-power-capacity-by-technology-2010-2027 (accessed August 27, 2023).

[7]   The Sun's Internal Structure and Atmosphere, Solar Wind - PMF IAS n.d. https://www.pmfias.com/sun-internal-structure-atmosphere/ (accessed June 29, 2023).

[8]   Overview | Sun – NASA Solar System Exploration n.d. https://solarsystem.nasa.gov/solar-system/sun/overview/ (accessed August 12, 2023).

[9]   Jean J, Brown PR, Jaffe RL, Buonassisi T, Bulović V. Pathways for solar photovoltaics. Energy and Environmental Science 2015;8:1200–19. https://doi.org/10.1039/C4EE04073B.

[10]   Panda-shaped solar plant begins operations in

N China - CGTN n.d. https://news.cgtn.com/ news/334d6a4e7a557a6333566d54/index.html (accessed August 10, 2023).

[11] World's Largest Concentrated Solar Power Plant is in Dubai – HELIOSCSP n.d. https://helioscsp.com/worlds-largest-concentrated-solar-power-plant/ (accessed August 10, 2023).

[12] Brier J, lia dwi jayanti. Renewables 2014 Global Status Report. REN21. (2014).http:/www.ren21.net Portals/0/ documents Resources/GSR/2014 GSR2014_full%20report_ low %20res.pdf 2020;21:1–9.

[13] B.Eteiba M, T. El Shenawy E, H. Shazly J, Z. Hafez A. A Photovoltaic (Cell, Module, Array) Simulation and Monitoring Model using MATLAB®/GUI Interface. International Journal of Computer Applications 2013;69:14–28. https:// doi.org/10.5120/11845-7579.

[14] NREL, First Solar Collaboration Enhance Thin-Film Solar Cells n.d. https://www.saurenergy.com/solar-energy-news/nrel-first-solar-collaboration-enhance-thin-film-solar-cells (accessed August 10, 2023).

[15] Building-Integrated Photovoltaics - green building to the next level | Reynaers n.d. https://www.reynaers.com/ inspiration/stories/products/building-integrated-photovoltaics-green-building-next-level (accessed August 10, 2023).

[16] IPCC — Intergovernmental Panel on Climate Change n.d. https:// www.ipcc.ch/ (accessed March 23, 2023).

[17] https://pixels.com/featured/1-parabolic-trough-solar-power-plant-philippe-psaila.html n.d.

[18] Https://5shopai.site/products.aspx?cname=linear +fresnel&cid=118. No Title n.d.

[19] Dish/Engine System Concentrating Solar-Thermal Power Basics | Department of Energy n.d. https:// www.energy.gov/eere/solar/dishengine-system-concentrating-solar-thermal-power-basics (accessed August 11, 2023).

[20] Dish type Stirling solar thermal power generation technology n.d. http://www.micropowers.com/en/

Dishstirling.aspx?cid=418 (accessed August 11, 2023).

[21]   Homepage - U.S. Energy Information Administration (EIA) n.d. https://www.eia.gov/ (accessed March 23, 2023).

[22]   Jacobson MZ, Delucchi MA. Providing all global energy with wind, water, and solar power, Part I: Technologies, energy resources, quantities and areas of infrastructure, and materials. Energy Policy 2011;39:1154–69. https://doi.org/10.1016/J.ENPOL.2010.11.040.

[23]   Energy Technology Perspectives 2014 – Analysis - IEA n.d. https://www.iea.org/reports/energy-technology-perspectives-2014 (accessed March 23, 2023).

[24]   Cook TR, Dogutan DK, Reece SY, Surendranath Y, Teets TS, Nocera DG. Solar energy supply and storage for the legacy and nonlegacy worlds. Chemical Reviews 2010;110:6474–502. https://doi.org/10.1021/CR100246C/ASSET/CR100246C.FP.PNG_V03.

---

[1] Biomass refers to organic material derived from plants or animals that can be utilized for energy production, such as wood, crop residues, food and yard wastes, animal manure, and algae.

[2] In the Sun's core, hydrogen nuclei (protons) undergo a series of fusion reactions to form helium nuclei. The most common fusion reaction in the Sun is the proton-proton chain reaction, which consists of several steps. In the first step, two protons combine to form a deuterium nucleus (a proton and a neutron) with the release of a positron (a positively charged electron, the antiparticle of the electron) and a neutrino. This reaction can be shown as follows:

(Proton + Proton → Deuterium + Positron + Neutrino)

The mass of two protons (primary particles) is slightly higher than that of a deuterium nucleus, positron, and neutrino (final particles). This mass difference is converted into energy according to Einstein's equation. The following steps in the proton-proton chain reaction involve the fusion of more deuterium nuclei to produce helium nuclei. At each stage, some mass is converted into energy. The general reaction can be summarized as follows:

(4Protons → Helium + 2 Positrons + 2 Neutrinos + Energy)

A small fraction of the initial mass is converted into energy during the fusion process. Positrons annihilate with electrons, resulting in additional radiation. The mass of the helium nucleus is 0.635% less than the mass of four protons, and this mass difference is converted into energy according to Einstein's equation. This energy is radiated as light and heat and provides the brightness and warmth of the Sun.

[3] One of the most valuable and efficient laws about radiation is the Stefan-Boltzmann law. Named after physicists Joseph Stefan and Ludwig Boltzmann, this law relates the total energy radiated by a perfect blackbody to its temperature. This law states that the total radiant power emitted per unit area of a black body is directly proportional to the fourth power of its absolute temperature. Mathematically, it can be expressed as follows:

$$P = \sigma A T^4$$

Where P represents the radiant power or the total energy emitted per unit time, A is the surface of the black body, T is the temperature of the black body in Kelvin, and σ is the Stefan-Boltzmann constant. This law shows that as the temperature of the black body increases, more energy is emitted exponentially and with the fourth power per unit area.

[4] Gross Domestic Product (GDP) is a measure of the total value of goods and services produced within the borders of a country in a specific period (usually a year). It includes the value of goods and services produced by domestic and foreign entities within the country's borders. In the field of solar energy, GDP can be used to evaluate the economic impact of solar energy on a country. For example, installing and maintaining solar energy systems can create jobs and contribute to a country's GDP. In addition, using solar energy can reduce dependence on fossil fuels, thereby helping to facilitate a country's trade deficit and increase its GDP.

[5] The most common solar cell types are those made of crystalline or conventional, traditional, first-generation, or wafer silicon. This type of solar cell is made of solar wafers with a thickness of 160 to 190 micrometers, which are slices of solar-grade silicon.

[6] A Circular economy is a regenerative system in which resources are used as long as possible, waste is minimized, and materials are continuously recycled.